Sex Dolls at Sea

Media Origins
Edited by Elizabeth Losh and Celia Pearce

Numbered Lives: Life and Death in Quantum Media, Jacqueline Wernimont, 2018
Sex Dolls at Sea: Imagined Histories of Sexual Technologies, Bo Ruberg, 2022

Sex Dolls at Sea

Imagined Histories of Sexual Technologies

Bo Ruberg

The MIT Press
Cambridge, Massachusetts
London, England

The MIT Press would like to thank the anonymous peer reviewers who provided comments on drafts of this book. The generous work of academic experts is essential for establishing the authority and quality of our publications. We acknowledge with gratitude the contributions of these otherwise uncredited readers.

This book was set in Stone Serif and Stone Sans by Jen Jackowitz. Printed and bound in the United States of America.

Library of Congress Cataloging-in-Publication Data

Names: Ruberg, Bonnie, 1985- author.
Title: Sex dolls at sea : imagined histories of sexual technologies / Bo Ruberg.
Description: Cambridge, MA : The MIT Press, 2022. | Series: Media origins |
    Includes bibliographical references and index.
Identifiers: LCCN 2021033923 | ISBN 9780262543675 (paperback)
Subjects: LCSH: Sex dolls—History. | Sex—Social aspects. | Sex (Psychology) |
    Sex in mass media.
Classification: LCC HQ23 .R83 2022 | DDC 306.77—dc23
LC record available at https://lccn.loc.gov/2021033923

10 9 8 7 6 5 4 3 2 1

For Jonah, who, like me, has always loved the water

Contents

Conclusion: Reclaiming the *Dames de Voyage*—The Feminist Potential of a Fictional Past   213

Series Foreword

Media Origins is a venue for interdisciplinary, humanistically informed research that recovers and interrogates the origin stories of contemporary media technologies. The titles address a range of cultural objects in the history and prehistory of computation. The series explores the politics of design and labor, the role of economics more broadly imagined, and the cultural frameworks of shared meaning making that undergird not only innovation but also maintenance, consumption, and disposal. Such origin stories often examine precomputational precursors to understand the larger social patterns, values, and beliefs behind a given medium's trajectory into the contemporary technological milieu. Volumes in the series may deploy feminist, postcolonial, queer, or antiracist theory to foster deeper conversations about the framing narratives of innovation.

The Media Origins series cautions that, in its obsession with the new, "new media" have developed an alarming ahistoricism that puts media studies at risk of losing valuable and largely undocumented accounts, particularly when cultural memory resides in rapidly aging witnesses or in records that are precariously stored in informal or neglected archives. Rather than reinforce assumptions about the technological survival of the fittest based on market metrics, the series excavates foundational platforms that have been all but ignored due to their perceived lack of commercial success.

Media Origins was launched to counter historical narratives that tend to emphasize the "inventor myth," crediting a lone auteur. Unfortunately, overtelling one origin story usually comes at the expense of often-marginalized groups and participants that were instrumental at inception or adoption. Equally damaging to understanding media origins can be the reification of artifacts with little attention to the larger discursive contexts of their invention, manufacture, and adoption. In looking at the interactions between actors and objects, books in the Media Origins series may revise existing views about the dynamics of power and control, specialization and distribution of labor practices, or systems of credit.

Acknowledgments

The process of researching and writing this book has often felt like a solitary one, the result of many months spent hiding away in my office, pouring over documents, delving deeply into archives, and thinking about the relationship between technology, sexuality, and this odd thing we call *history*. In reality, this project has come into being through the generosity and support of many people—in no small part because much of my work has taken place during the COVID-19 pandemic. Some of these people have been my collaborators, colleagues, and friends. Others have been archivists and librarians. A number of others have lent me their assistance even though we have no official professional ties. For a number of years now, I have been that stranger on the internet sending cold-call emails about sex dolls to anyone I had reason to believe might hold a clue I could use in my search for the *dames de voyage*. To my great benefit, almost all of them answered, and I learned something new from each one.

A number of the scholars whose work I discuss at length in the early chapters of this book have generously replied to my queries about their sources and shared with me some of their unpublished works. Among these are Cynde Moya, Anthony Ferguson, David Levy, Amy Wolf, and Hallie Lieberman. I am also grateful to Minsoo Kang for his thoughts on the veracity of the tale of the dames de voyage, as well as to scholars and archivists of maritime history Gina Bardi, Peter Kasin, Gibb Schreffler, and Amy Parsons for their expert guidance. My thanks to Andrea Horbinsky and Zoyander Street for their assistance with my research into the textual lineage behind the story of the "Dutch wives," specifically in verifying my readings of Japanese-language texts. Relatedly, thank you to Kathryn Levine and Aubrey Gabel for indulging me and reading a schlocky serialized short story from the 1890s so that we could double- and triple-check the context meaning of references to dames de voyage.

Thank you to Elizabeth Losh and Celia Pearce, the editors of the Media Origins series, for believing in this project even in its nascent stages and for supporting my vision of a book that is somehow simultaneously quirky and serious, highly specific and extremely wide-ranging. My gratitude also goes out to my editors at the MIT Press, first Doug Sery and now Noah Springer, for their encouragement and support through the publishing process. I am thankful for the work of my anonymous peer reviewers, who offered feedback at multiple stages, including during the hubbub of the pandemic. Their feedback on the original full draft of the manuscript, which was admittedly still rough, was invaluable. I am especially grateful to my most exacting reviewer, whose careful engagement with the manuscript draft was exactly the push I needed to make this into the book it is today.

Librarians, archivists, and curators are the true heroes of this project. My enormous thanks to Jenna Dufour, research librarian for the visual arts at the University of California, Irvine. I have lost track of the number of times I have emailed Jenna in a state of exuberance or panic to ask for help tracking down some obscure sex-related document. The Humanities and Rare Books Reference Teams at the British Library, and Elias Mazzucco in particular, went above and beyond to help me track down the album of advertisements for sex toys discussed in chapters 2 and 4. Despite a series of obstacles, from pandemic travel restrictions to the realization that the documents in question were too fragile for reproduction or public viewing, the British Library staff was able to send photos at the eleventh hour that I was so happy to receive that I spent a whole morning alone in my office laughing with delight.

My thanks also to the staff at both the ONE National Gay and Lesbian Archives and the library at the University of California, Los Angeles, who valiantly ventured into the stacks to produce scans for me during pandemic lockdown. Whoever it is at the Internet Archive who decided it was worthwhile to preserve early rubber goods catalogs and minor works of German sexology in a meticulous array of published editions, you have contributed to this research in more ways than I am sure you will ever know. Many of the primary texts in this book, and especially those that reflect French popular print culture from roughly 1850 to 1920, come from the amazing digital archive Gallica, associated with the Bibliothèque Nationale de France.

Additional archives and collections that I have drawn from include the American Periodicals Series database, the holdings of the British National Maritime Museum, and a wide range of print texts flung to the far corners of the American interlibrary loan system. To all who have worked to create and maintain those collections, my thanks.

Many colleagues and students have been instrumental in helping me develop my thinking about sex dolls, sex tech, and the history of sexual technologies. Thank you to Josef Nguyen for the chance to think alongside his developing work on sex dolls and consent. Thank you also to Bliss Cua Lim for suggestions of work that bridge the sex doll and broader artistic traditions exploring doll imagery. A special thank you to all of the members of the Critical Approaches to Technology and the Social (CATS) lab at the University of California, Irvine, which I co-run with Aaron Trammell: Ryan Rose Aceae, Kat Brewster, Amanda Cullen, Will Dunkel, Nazely Hartoonian, Ke Jing, Ian Larson, Rainforest Scully-Blaker, Bryan Truit, Isabelle Williams, and Cass Zegura. You are the best group of graduate students a proud faculty mentor could ask for, and your feedback has invaluably shaped my revision of this project. Kat Brewster is also the artist who created the original paintings of early sex toys found at a number of points in the book. Thank you, Kat, for your amazing work, which has helped bring this history to life.

I am grateful to others in my academic community for their support and guidance. Thank you to my colleagues in the Department of Film and Media Studies and the School of Humanities at the University of California, Irvine, whose warmth and enthusiasm for my work has been truly sustaining. Thank you to my department chair, Fatimah Tobing Rony, and to Dean Tyrus Miller for offering me a course release to make up for research time lost during the pandemic, which served as a major factor in allowing me to complete the revision of this project on time. Thank you to the UCI Humanities Center for awarding me a publication grant to assist with the cost of indexing and proofreading this project. Beyond UCI, thank you to Jacque Wernimont, whose book *Numbered Lives* precedes mine in the Media Origins series and who has been a role model in the work of writing feminist media histories.

In a sense, this book represents a coming together of numerous elements of my life that go back many years. For that reason, all of the people I should truly thank are too many to list here. I will, however, say thank you to the Comparative Literature program at the University of California,

Berkeley, where I earned my PhD. My mentors there taught me to close read the world around me as text. They also created an environment that fostered the foreign language skills I would later need to undertake such studious endeavors as discerning the difference between a sex doll made by a sailor and one made by a prisoner, or combing through old directories of the city of Paris to map a network of turn-of-the-century sex toy sellers. Thank you to those friends who have helped me grow my collection of antique vibrators over the years. Having firsthand knowledge of these material objects has helped me in ways I never anticipated when scouring estate sales for crumbling old sex toys tucked away in the back of closets.

I love writing, and I have loved writing this book. To be honest, though, I have also found it to be the hardest thing I have ever written. Often, working on it has felt like sitting at the bottom of a swimming pool, fully submerged, with the world above somehow very far away. Throughout this process, the people closest to me kept me grounded and reminded me that, though writing is my first love, there are many other things in the world worth loving. Thank you to Eli Peterson, who has listened to me with an exceptional amount of patience as I have rambled on about the minutiae of this project. His support, in both love and time, is absolutely what has made this book possible. Thank you also to Jonah Peterson. He cannot read yet but hopefully one day, years down the line, he will pick up this book and realize that this strange volume—with its sex dolls and its sailors and its dreams of remaking history—is what I was working on all those evenings he sat beside me and kept me company while I was writing.

Introduction: "The Beginning of the Modern Sex Doll"— Imagining the History of Sex Tech

Sex dolls are having a moment. They are more and more the subject of news stories, whether laudatory or mocking, about the rise of increasingly "realistic" devices for having sex through or with technology.[1] Although there are many different types of devices that fall into the category of sexual technologies, the sex doll—along with its high-tech counterpart, the sex robot—sits at the center of this growing public interest. In America, companies like Real Doll, which makes elaborate, customizable life-sized sex dolls, have drawn particular attention, but there is a whole wider world of sex dolls and related items currently available on the consumer market.[2] Today, it is possible to purchase everything from exceedingly busty anime "love dolls" to torso sex dolls whose bodies extend only from their breasts to their genitals to the iconic Fleshlight, a "pocket pussy" with an opening for insertion that can be shaped like a vulva, anus, or mouth. The presence of sex dolls in our pop cultural imaginaries is growing just as rapidly as— or, likely, more rapidly than—the development and use of sex dolls themselves. Sex dolls have made appearances in works of literature and other media for centuries. Now, as those speculative visions seem to be on the brink of becoming realities (at least according to eager roboticists and sex toy manufacturers), the presence of sex dolls in mainstream reporting and academic research alike has spiked. This is especially true in contexts related to technology, where both fantasies and anxieties about the future of computation are increasingly entangled with fantasies and anxieties about sex.

However, sex dolls are not just a thing of the present or even the future. They are also very much a thing of the past. Like sex dolls, dolls more broadly have been a point of fascination across cultural and national contexts; they feature in long-standing lineages of fiction, art, film, and new media, in which the figure of the doll is almost always highly gendered and eroticized, frequently blurring the divide between *doll* and *sex doll*.[3] Yet the history of the sex doll goes beyond fiction, into the material realities of sex dolls themselves, as the research presented in this book makes clear. Sex dolls and other sexual technologies have had their culture "moment" multiple times before. They have been an integral part of earlier waves of technological innovation and former iterations of what we might now term the *sex tech craze*. From roughly the 1850s to the present, from the popularization of vulcanized rubber in the mid-nineteenth century to the sexual revolution of the 1960s and 1970s and up to the technomasculine maker cultures of the last two decades, sex dolls have been a recurring point of interest and debate—not just in more rarified realms of literature and art but also in the industries, consumer markets, and mainstream cultural conversations surrounding technology.

Others before me have written more traditional versions of the sex doll's history. Indeed, many of these will return as objects of analysis in this book.[4] What readers will find here, by contrast, is not primarily a retelling of the history of the sex doll—though there is much to learn about that history in the chapters that follow. Instead, first and foremost, the present work is an interrogation of how the history of sex dolls itself has been imagined: how established narratives about that history have come into being, whom they have privileged and whom they have marginalized, how we can reimagine the history of sex dolls, and how this act of reimagining the past might itself represent a crucial step toward creating a more socially just future for sexual technologies.

In particular, I am interested in the sex doll's origin stories. These are the tales that are told (and retold) about where the sex doll came from and who made the very first sex doll. As we will see, though these origin stories are ostensibly about sex dolls specifically, they are often used as starting points for broader histories of sexual technologies, positioning the sex doll as the precursor to all of the various sexual devices that have followed or will follow. How we tell the story of the origins of the sex doll sets the stakes for how we tell the history of sexual technologies on a broader scale. And, in turn, how we tell the history of sexual technologies forms a baseline for implicit assumptions about which people and which desires matter when it comes to the intersection of sex and tech. These origin stories are often quirky and seemingly harmless, adding a bit of color and relatability to narratives about the evolution of sexual devices that are still commonly viewed as uncanny, unnatural, or unethical.[5] Yet when examined closely, these origin stories become windows into much larger issues of power and identity that shape the cultural landscape of tech today—a landscape where celebrations of technological "advancements" often thinly veil discriminatory attitudes toward women, queer and transgender people, and people of color.[6] The origin stories of the sex doll are the origin stories of technology. To change the course of technology, we need to change the history of the sex doll.

Of all the possible origin stories for sexual technologies, this book focuses on one in particular: the tale of the dames de voyage. The tale of the dames de voyage ("women of travel" or, more loosely, "traveling companions"), which appears in numerous contemporary accounts, claims that the very first sex dolls were rudimentary figures cobbled together out of cloth and leather scraps by European sailors on long, lonely voyages in centuries gone by (figure 0.1). These makeshift dolls were reportedly filled with cotton or

Figure 0.1
The French naval ship *Ville de Paris*, completed in 1764, is an example of the type of ship on which contemporary historians say that sailors made their own sex dolls.

straw, presumably with holes cut between their legs for penetration, and shared between sailors. If the details of this account sound murky, that is because the story exists today in multiple forms, many of them contradictory. Various authors state that the sailors' dolls originated as early as the 1600s or as late as the 1900s; they credit the dolls' invention to sailors from an array of European countries, from France to the Netherlands to England to Spain. A number of these authors accompany their iteration of the tale with what they understand to be a photograph of surviving examples of the dames de voyage, visual proof of the dolls' existence (figure 0.2). One thing that all of these accounts (and there are many of them, so many that each time I sit down to work on this book I find that new ones have been published) have in common is that they take the story of the sailors' sex doll to be true. This vision of the very first sex doll, inspired by the longing of lusty men at sea, is presented as an indisputable and indeed pivotal moment in the history of sex tech.

Questioning this unquestioned origin story is the central impetus behind this book. Along with that comes the need to make sense of the answers that such a line of questioning reveals. To accomplish this, I use a multipronged set of methods, combining a Foucauldian approach to tracing genealogies with the tools of intersectional feminist analysis.[7] I begin in the present, mapping contemporary iterations of the tale of the dames de voyage. From there, I move backward to chart the lineage of writing that has brought the story of the sailors' dolls into the present day. In searching for the dames de voyage, I move through an eclectic array of archives, interweaving primary and secondary documents from across historical and intellectual moments. These range from works of sexology and sociological pseudoscience to cultural ephemera: songs about the wonders of rubber, advertisements for inflatable vaginas, and news reports about bootleggers transporting illegal alcohol inside sex dolls, to name only a few. Having explored these histories (both the history of the sex doll and the history *of the history* of the sex doll), I embark on an interrogation of the tale of the dames de voyage itself as an origin story. Through this analysis, I lay bare a web of concerns relating to gender, sexuality, race, and colonialism that lie just beneath the tale's surface, making visible the vital importance of asking: Why are certain origin stories about sexual technologies told over others? What cultural work are these stories deployed to do? Whom do these stories serve and whom do they erase?

Koitus-Ersatz
Entkleidbare, mit allen weiblichen Kleidungsstücken ausgestattete primitive Puppe fast in
Lebensgröße
(Archiv des Instituts für Sexualwissenschaft, Dr. Magnus Hirschfeld-Stiftung, Berlin)

Tab. XX   Zur Sittengeschichte des Pygmalionismus

Figure 0.2
Image commonly reprinted in contemporary histories and labeled as *dames de voyage* or *sailors' sex dolls. Source:* Leo Schidrowitz, *Ergänzungswerk zur Sittengeschichte des Lasters* (Supplement to the moral history of vice) (Vienna: Verlag für Kulturforschung, 1927).

This book began, roughly four years ago now, with a moment of curiosity. I was teaching a graduate course on the relationship between sexuality and technology, and, in one of our texts, I found myself reading an anecdote that I had read many times before: the tale of the dames de voyage. The story had always intrigued me, and it still does. Despite all the objections I have to it—and you will find that this book is full of them—there is something alluring in the vision of the very first sex doll made and used at sea. Perhaps because I felt more accountable for this story as a teacher than I had previously as a reader, I set out to learn more about these fabled sailors' dolls. What were they like? Did they even exist? What I thought would be a brief dive into existing research quickly revealed itself to be something much more. In trying to answer these questions, I found myself uncovering a wide-reaching tangle of histories, some real and some imagined. As the pieces fell into place, they began to tell a larger story, one that starts with sex dolls but has relevance that extends into many more areas. It is a story about how history is made as much through fantasy as through fact. It is a story about how the history of technology cannot be separated from the history of sex. Above all, it is a story about how the history of sexual technologies as we think we know it can be destabilized and reclaimed and how sex tech itself can be made anew through the radical power of speculation.

Sexual Technologies, Historical Imaginaries

Sex and technology are fundamentally intertwined.[8] This is true across the past, the present, and the future. Today, in the early decades of the twenty-first century, it has become a cliché to say that sex is ever more technological. In America, the cultural context from which I write this book, digital tools now facilitate and mediate many aspects of our sexual lives. Smartphone users find romance and sex via dating apps; social media offers platforms for expressing sexual identity; sex parties are being hosted over web conferencing software like Zoom.[9] Yet long before sexting or webcam modeling or internet-enabled sex toys, technology (broadly defined) was already central to sex, and sex was already central to technology.[10] We know, of course, that the popularization of many new consumer technologies and media forms has been tied to the sale of pornography, as stories about the rise of the printing press, the VHS tape, and the internet all make clear.[11] However, technologies are more than vessels for transmitting sexual content. They are also themselves tools for having sex and even having sex *with*. There is no piece of technology that has never been brought into the bedroom. There is no medium of communication—from the SMS to the telegraph to the carrier pigeon—that people have not used to have sex (also broadly defined) with one another.[12] No matter how low-tech it may seem, sex itself is always bound up with technology. Contraception, legalization, and reliable access to private spaces are all, in their own ways, "technologies" that directly shape the sexual practices of individuals who may never send a nude picture online or purchase a sex toy.[13]

At the intersection of sex and technology stand what I refer to as *sexual technologies*. By this I mean something both specific and broad. In the most straightforward sense, sexual technologies are devices designed for sexual uses. Some of these are built around what we think of as twenty-first-century technologies, like the web connectivity of teledildonics (sex toys that hook up to computers) or the artificial intelligence of sex robots.[14] Other examples come from a predigital time. These use what we think of as technologies of ages past: pneumatic tubes, mechanized gears, elastic materials that are banal today but were once technological marvels. It is also important to note that there is not such a clear divide as we typically imagine between the sexual technologies of the past and the sexual technologies of the present. The "fucking machines" of the 2000s, which embodied the DIY erotics of the sex toy hacking scene, look nearly identical to "women's masturbation machines" being produced in Germany by the 1920s (figure 0.3).[15] A century before meticulously designed vibrators were winning innovation awards at major electronics expos for their promises to deliver "the holy grail of orgasms," dildos that could suck up and release warm water were being advertised under nearly identical promises: to offer new heights of sexual pleasure and make both women and men orgasm so intensely that they would "bawl with happiness."[16] The category of sexual technologies also extends beyond technologies explicitly designed for use during sex. It includes a wide range of technologies that have facilitated shifts in sexual practices, including technologies of governance and oppression. This vision of sexual technologies is intentionally both capacious and ambivalent.

Throughout this book I also refer to the concept of *sex tech*. To an extent, sex tech is a shorthand for sexual technologies, but the two are not quite synonymous. *Sex tech* is a predominantly twenty-first-century term that

Onaniermaschine für Frauen
Gewerbsmäßig hergestellter Kohabitationsapparat, der polizeilich beschlagnahmt wurde und
von dem sich ein Original im Dresdner Kriminal-Museum befindet
(Archiv des Instituts für Sexualwissenschaft, Dr. Magnus Hirschfeld-Stiftung, Berlin)

Figure 0.3
A masturbation machine for women dating from the 1920s or earlier, which resembles
twenty-first-century "fucking machines." *Source:* Leo Schidrowitz, *Ergänzungswerk zur
Sittengeschichte des Lasters* (Supplement to the moral history of vice) (Vienna: Verlag
für Kulturforschung, 1927).

signifies, as I use it, a set of contemporary cultural conversations around
sexual technologies. Although the reach of sex tech is now wider and more
diffuse, in its origins sex tech is inextricable from the 1990s and 2000s
tech cultures that took root in places like the San Francisco Bay Area.[17]
At the core of this tech culture was (and, to an extent, still is) a vision of
a postidentity, postpolitical society in which those who ride the cutting
edge of technology are also, not coincidentally, those who have the best
sex.[18] Sex tech today is also closely tied to both the adult entertainment
and technology industries. Events such as the Adult Entertainment Expo

have been central locales for promoting increasingly "innovative" sexual technologies.[19] Major online porn-sharing platforms have been involved in pushes for the development of sexually explicit big-budget video games and sex-based virtual reality experiences.[20] At the same time, marginalized creators are operating under the umbrella of sex tech in order to experiment with repurposing standard technologies to create opportunities for queer play.[21] At the end of the day, more than any particular set of technologies or practices, sex tech is an idea: a vision of how sex could be sexier and tech could be "techier" by increasing our erotic entanglements with machines. Above all, as Lynn Comella has compellingly argued, sex tech is a cultural imaginary, a set of shared notions about what sexual technologies are, have been, and could be.[22]

Sex tech has also played an important role in my own life. For the five years before beginning graduate school, when I would embark on my career as a scholar of gender and sexuality in digital media, I worked as a sex tech journalist. This was roughly 2005 to 2010, and at that time sex tech meant something scrappy, upstart, and seemingly full of potential. These were the years of user-crafted genitals in *Second Life*, of the rise of Kink.com—a time when sex workers were making news by opening their own websites streaming live video of themselves in their bathrooms. For much of this period, I was living in San Francisco, where sex tech seemed to be all around me: in hardwired butt plugs on display at Arse Elektronika, in the crossovers between the local porn scene and the local tech scene, and in the sci-fi erotica read out by fellow writers at the now sadly defunct Center of Sex and Culture.[23] Sex tech in San Francisco was, in retrospect, a problematic scene: overwhelmingly white and surprisingly straight, and perhaps a little too eager to induct a twenty-three-year-old then-femme-presenting journalist into the fold. For better or for worse, though, it is through sex tech that I got my entry into the work I do today, which has since taken a far more critical and theoretical turn. For this reason, I am both invested in the sex tech imaginary and highly skeptical of it. I recognize its appeal, but I also have the perspective now to see how this imaginary is structured around deeply biased notions of who and what sexual technologies are for.

I have also come to realize, through writing this book, that the sex tech imaginary has not only formed through notions of what sexual technologies are today or what they could be tomorrow. That imaginary has also been built around—and continues to be sustained through—the telling of certain versions of history. Although the spirit of sex tech seems fundamentally future-facing, the sex tech imaginary is also a historical imaginary. In part, that is because many of the contemporary histories that have been written about sex tech are imaginary in the most basic sense: they are erroneous, made up, imagined. That is simultaneously surprising and unsurprising. The work of writing both the history of sexuality and the history of technology come with similar challenges (incomplete records, obsolescence of one kind or another, the problem of making meaning from traces of everyday life that exceed the bounds of the archives). These challenges are amplified when the two subjects are brought together. Sexual histories in particular are vulnerable to misguided essentialisms that, as Gayle Rubin explains in "Thinking Sex," "consider sex to be eternally unchanging, asocial, transhistorical." From this viewpoint, writes Rubin, "sexuality has no history."[24] Certainly, following the work of Michel Foucault, historians of sex know better than to think in such essentialist ways.[25] However, the authors whose histories I most stringently critique here are not historians of sex; they are researchers of technology, or technologists themselves. Among those whose expertise lies first and foremost in tech, it often seems that sexuality still has no history. Sexual desire—and straight, cisgender, white men's desire in particular—continues to be imagined as timeless, a force so absolute that it can be used to explain technological innovations from the present day back to the dawn of time.

In truth, whether accurate or erroneous, all histories of sexual technologies are imagined histories. Each way of envisioning the past entails a set of decisions about how to imagine what has come before. This is true for the history of sex tech. It is true for the history of sex dolls. It is also true for the history of technology and the history of sexuality more generally. When it comes to technology and sex, the politics of the present lie in the practices of envisioning the past—regardless of what is true.

Telling Stories about the Origins of Sex Tech

There are many ways to tell the history of sexual technologies, and just as many individual devices or cultural phenomena that we could point to as the "origin" of sex tech. While some of these possible origin stories may be more historically correct than others, each comes with its own set of cultural implications. In effect, each paints a different picture of which people, which bodies, and which pleasures have been central to the development of sexual technologies. Each imagines a specific starting point, a specific moment of inception that gives birth to all of the sexual technologies to come.

Outside of the realm of the sex tech buzz, a number of histories of sexual technologies have focused on devices designed for women's bodies, such as dildos and vibrators.[26] From a feminist perspective, the dildo seems like a promising origin point for sex tech, since it suggests that a long lineage of sexual technologies has emerged from an item developed for women's pleasure. However, kicking off this history with the dildo also raises issues. Very few historical accounts of the dildo sufficiently address the use of insertable objects by men or the use of dildos for sex between women (something that was very much on the minds of commercial dildo manufacturers as early as the start of the 1900s). Nor does it address the blurry line between dildos and what we would now call *packers*: nonrigid, penis-shaped objects often worn by transmasculine people in contexts outside of sex.[27] Descriptions of the dildo as an "ancient" technology also often exoticize nonwhite, non-Western peoples, using references to the dildo's supposed development in societies in Asia and Africa to communicate a sense that sex toys have their roots in "primitive" sexual traditions.[28] The case of the dildo illustrates how the same origin story simultaneously can be counterhegemonic, contesting masculinist visions of technology's history, and also can perpetuate the marginalization of queer and trans experiences, as well as the experiences of people of color and those from the Global South.

Another way of telling the origins of sexual technologies is through a focus on the vibrator. However, the vibrator comes with pros and cons of its own. Like the dildo, it has the advantage of centering women in the history of sexual technologies, in some cases drawing attention to women as entrepreneurs in the field of sex toys.[29] The vibrator also makes ties between sexual history and technological history usefully apparent, given that the vibrator is likely to be more readily interpreted as a form of "technology" than the dildo. Yet the vibrator itself has its origins in the medicalization and pathologization of women's bodies, dating back to a time when it was used by doctors to control women's sexuality through the treatment of so-called hysteria, as Rachel Maines documents in her foundational book, *The Technology of the Orgasm*.[30] These roots have unexpected echoes in contemporary sex tech—for example, in the form of vibrators and other "smart"

sex toys that can be manipulated by one's partner at a distance.[31] Perhaps
the real value in looking to the vibrator as the origin point for sexual technologies, then, lies in bringing to the fore how the development of such
devices has always been as much about control as it is about pleasure.

The sex doll itself, as one possible origin point for sexual technologies,
has many possible origin stories of its own. One of these is the tale of the
dames de voyage, which credits European sailors from somewhere between
the 1600s to the 1900s with the invention of the sex doll, as I explain at
length in chapter 1. However, there are also many other origin stories for
the sex doll that run alongside or stand in place of the tale of the dames de
voyage. Take, for example, the myth of Pygmalion, references to which have
recurred in writing about sex dolls for at least the last two hundred years.
Pygmalion, as he appears in Ovid's *Metamorphoses* and many other works,
is a sculptor who carves a statue of a beautiful woman and then promptly
falls in love with her.[32] Later, the goddess Aphrodite brings the sculpture to
life, and she becomes Pygmalion's bride. As I discuss in chapter 7, the myth
of Pygmalion makes for a curious origin story. Rooting sex tech in classical
myth gives the later lineage of sexual technologies an aura of gravitas. It
also suggests that those who create sex dolls are, like Pygmalion, artisans
of mythical skill. Yet if we look back to the actual text of *Metamorphoses*,
we find that Pygmalion is portrayed less as an awe-inspiring inventor and
more as an awkward, oversexed romantic who hates women and idealizes
the not-quite-liveness of dolls. The myth of Pygmalion, precisely because
it is a myth, helpfully literalizes the fictional natural of the sex doll's various origin stories. It also exemplifies a trend that recurs across such origin
stories: that the same tales that are supposed to make sexual technologies
seem legitimate, natural, and manly end up making them seem precarious,
comedic, and queer.

Other origin stories for the sex doll place its creation in an array of different time periods and political contexts. According to another alternate
tale that I also discuss in chapter 7, sex dolls were invented by Nazis during
World War II for use by German soldiers. Still other stories look outside of
Western Europe for the sex doll's origins, typically to Japan or Asian countries more generally, such as in the story of the "Dutch wives," which I
address in chapter 6. Historian Agnès Giard has documented the popularity
of contemporary narratives (mostly emerging from Japan itself) that claim
that "love dolls" were first invented and produced during the Japanese Edo

period—a notion that Giard debunks, calling it a reflection of a "collective phantasmagoria" entirely lacking in material evidence.[33] Additional origin stories exist for the twin figure of the sex doll: the "gynoid" automaton, a robot shaped like a woman who has featured in many prominent works of literature and film across the nineteenth and twentieth centuries, from E. T. A. Hoffmann's 1818 "The Sandman" to Auguste Villiers de l'Isle-Adam's 1886 *L'Ève future* to Fritz Lang's 1927 *Metropolis*.[34] Relatedly, Minsoo Kang has documented what he terms the *intellectual fable* of René Descartes's mechanical daughter—a widely circulated yet fabricated story about an automaton shaped like a young girl supposedly created by Descartes, whom the philosopher took onboard a ship, only for her to be thrown overboard by the ship's captain. With its focus on a doll-like object on a journey at sea, this story has much in common with the tale of the dames de voyage, further drawing out its technological valences.[35]

Among these origin stories for the sex doll, the tale of the dames de voyage is arguably the most prominent today, and its prominence is snowballing. The tale has appeared in numerous works, both popular and academic, that are proving to have a long-lasting influence on the ways that subsequent authors understand the history of the sex doll and the history of sexual technologies. However, before homing in on the tale of the dames de voyage, it is valuable to reflect on these other imagined origins. They illustrate how the story of the sailors' dolls exists within an ecosystem of such stories, each of which evolves through its own textual lineage and each of which offers a different way to think about the cultural work that origin stories set out to accomplish. While finding out the real history behind the tale of the dames de voyage is one of the goals of this book, my overall purpose is not to whittle down this network of origin stories to identify the "right" one. Writing in *Imperial Leather*, Anne McClintock describes how interrogating past sexual practices offers a chance to explore "complex, historically diverse phenomen[a] that cannot be reduced to a single, male, sexual narrative of origins."[36] The result of such an exploration "is less an attempt to empirically recover the past than it is an attempt to intervene strategically in historical narratives," says McClintock, "in such a way as to throw into question not only the historical force of these relations . . . but also their continuing implications of our time."[37] Dominant narratives about the history of sexual technologies require precisely this sort of intervention, one that throws the past into question and by extension argues for new ways of thinking, making, and being in the present.

A History at Sea: Setting Sail in Search of the "Very First Sex Doll"

What is the reality behind the tale of the dames de voyage, and what are the implications of the tale as it is being told today? This book tackles these questions in two sections. Part I, Searching for the Very First Sex Doll, is about moving backward from the present day to find the fabled sailors' sex dolls. It begins with chapter 1, "Contemporary Tales of the *Dames de Voyage*: The History of an Imagined History." There, I map out various versions of the tale from across a wide web of twenty-first century sources, with a focus on those that explicitly position the dames de voyage as "the beginning of the modern sex doll."[38] After charting similarities and differences across these renditions (where and when they supposedly take place, how they contextualize the story of the sailors' dolls within larger narratives), I move into a mode of scholarly sleuthing, tracing what I refer to as the *citational lineage* behind the tale of the dames de voyage. This is a genealogy of works that begins with twenty-first-century histories and runs back through microgenerations of English- and French-language sources. The process of tracing this lineage reveals that the story of the sailors' sex dolls, as it appears today, has in fact been formed through a series of slippages, amalgamations, and deliberate obfuscations closely tied to the gender politics of citation and attempts to make scholarship seem original and legitimate.

Yet the process of imagining the dames de voyage began long before the present day. In chapter 2, "How Fantasy Became History: The *Dames de Voyage* in Pseudoscience, Erotica, and Advertising," I leave behind histories written over the last few decades and instead track down a set of earlier texts that are often used to bolster claims about the dames de voyage—either that they were rudimentary sailors' dolls or elaborate mechanized proto sex robots (and sometimes, oddly, both). These texts include "scientific" books on sex and sexual devices published in America in the late 1960s and early 1970s, the same period when now-iconic midcentury blow-up dolls began to appear for sale in men's magazines in the United States, Britain, and Australia. They also include German sexological treatises from the 1900s to 1920s, French erotica from the 1890s through the 1900s, and advertisements for sex toys sold in Paris at the turn of the twentieth century (figure 0.4). These too form a lineage of works that construct an earlier vision of the dames de voyage. Yet read more closely, these works quickly

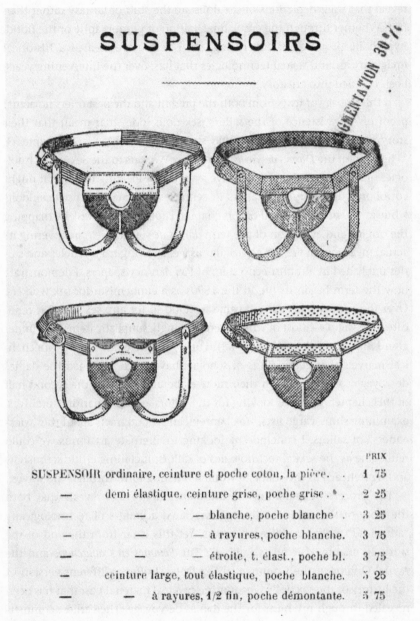

Figure 0.4
An advertisement for strap-on harnesses (euphemistically labeled *suspensoirs*) for use with dildos, offered for sale in 1900 by Parisian merchant Maison L. Bador.

reveal that their depictions of sex dolls are the stuff of fantasy rather than fact. Whether through the erotic imagination of pornography or the florid, hyperbolic promises of advertising, such texts have created a historical imaginary around sexual technologies that has, over the intervening years, been codified into "history."

If this lineage of texts from both the present and the past offers no actual proof of the existence of the sailors' sex dolls, does that mean that their story is entirely fictional? Or could such proof lie elsewhere? Chapter 3, "The Birth of the *Dames de Voyage*: From Sex Workers to the Sexual Technologies of Sailors," explores alternative avenues for locating traces that might corroborate the story that sailors in centuries past made or used sex dolls while at sea. In the first half of this chapter, I approach this task by mapping the origins and evolution of the term *dames de voyage* itself, uncovering its actual meanings in French vernacular usage. Drawing from articles and stories published in a robust ecosystem of Parisian newspapers, I demonstrate how the term began its life in the 1890s as a euphemism for sex workers. Over the next two decades, it came to stand in for any sex toy that replicated the vagina—most of which were strikingly simplistic items sold under grand marketing claims (figure 0.5). Thus, they were far from the mechanical marvels that generations of scholars have since imagined the dames de voyage to be, at least in their more elaborate forms. In the second half of this chapter, I shift to looking for the sailors' dolls in maritime archives, examining ships' cargo lists, sea chanteys, and moral tracts about the "wickedness" of sailors. I conclude by looking to alternate art forms we could reimagine as the sexual technologies of sailors, including erotic scrimshaw and figureheads in the shape of women bursting forth through the waves.

The process of searching for the dames de voyage reveals that both the story of the sailors' dolls and associated accounts of technologically "advanced" early sex dolls are myths. Yet this is far from the end of the story. In chapter 4, "'All Is Rubber!': The *Femmes en Caoutchouc* and the Actual Origins of the Commercial Sex Doll," I offer a different version of the history of the sex doll: a recounting of a real material past that has been revealed through my hunt for the dames de voyage. The earliest commercial sex dolls produced in Western Europe (and possibly the United States) were in fact the *femmes en caoutchouc*: "women" made of inflatable vulcanized rubber. These rubber women, which were one among many sex toys

Protective Women's Secret Part

No. 1   No. 2

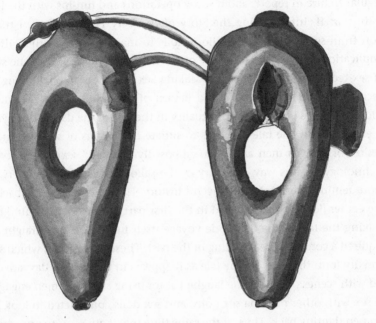

Figure 0.5
Artist's rendering of an inflatable rubber vulva of the sort historically referred to as a *dame de voyage* and advertised with grand marketing claims. Based on an image of a 1940 sex toy catalog from a Japanese seller that appeared in Paul Tabori, *The Humor and Technology of Sex* (New York: Julian Press, 1969). *Source:* Original art by Kathryn Brewster.

and other sexual devices to enter the market during a rubber boom that started in the 1850s, were most likely sold as a set of dismembered parts that could be folded up to fit in a box resembling a suitcase—a real-life manifestation of the imagery of disassembly and dismemberment that features in much literature and art related to dolls.[39] These parts could then be filled with air or water to give them a supposedly lifelike feel. By looking at advertisements, news reports, and fiction from the period, much of which circulated around the 1800s Paris World's Fairs, we see how the femmes en caoutchouc went through their own rise and fall in French popular culture.

Just as in the present, visions of the sex doll during the second half of the nineteenth century reflected a mix of awe and humor, with the rubber women eventually shifting from a symbol of technological innovation to a regular fixture in reports about seedy operations and run-ins with the law.

In part II, Interrogating the Story of the Very First Sex Doll, I transition from searching for the sailors' dolls to making sense of the cultural implications that underlie the use of the dames de voyage to tell the story of sex tech's history. Chapter 5, "Making Sex Tech Masculine, Making Sex Tech Straight: The Disavowal and Return of Femininity and Queerness," addresses issues of gender and sexuality in the tale of the dames de voyage. It explains how the tale attempts to situate the origins of sex tech in the desires of straight men and their supposedly "natural" sexual affinity for technology. In this way, the story of the sailors' dolls works to overwrite more feminist versions of sex tech's history. However, if we look back to the earlier documents described in the first part of this book, we find that making the tale of the dames de voyage itself masculine and straight has required a considerable rewriting of the past. These documents, which supposedly form the basis for the tale as it appears in the present day, are dotted with scenes of men being laughed at for using sex dolls, men engaging in sex with other men via sex dolls, and sex dolls constructed to look like women fighting back. Thus, at the same time that it attempts to undermine the contributions of women and queer people to the development of sexual technologies, the tale of the dames de voyage ends up painting a picture of sex tech's origins that is marked by femininity and queerness.

Issues of colonialism and racism have also shaped the historical imaginaries surrounding sexual technologies. In chapter 6, "From Bamboo Lovers to Undersea Kingdoms: Colonialism and Race in Stories of Sailors' Sex Dolls," I open by describing an alternate origin story for the sex doll: the story of the Dutch wives. The Dutch wives are imagined to have been early sex dolls used by sailors associated with the Dutch East India Company in the 1600s or 1700s, modeled off of bamboo sleep cages designed to keep sleepers cool on hot nights, and either created by the Dutch and adopted in Asia or created in Asia and adopted by the Dutch, depending on the account. The explicit ties between the Dutch wives, European imperialism, and orientalism prompt us to confront the colonialist and racialized logics that underlie the tale of the dames de voyage as well. Colonial violence facilitated the rise of rubber goods in the second half of the 1800s. Indeed, if we return once again to earlier documentation, we find that visions of white European sexuality went hand-in-hand with anti-Blackness and an erotic panic about colonial subjects. In its later sections, this chapter addresses echoes between the story of the sailors' dolls and the real experiences of enslaved people during the Middle Passage, arguing that the tale of the dames de voyage as it has been formulated draws imagery from this history while problematically attempting to distance itself from race. Drawing guidance from work by Afrofuturist thinkers who have reimagined the lives of those lost in the Atlantic slave trade, I ask whether the tale of the dames de voyage too can be reimagined through the power of speculation.

Just as revealing as the stories that are repeated about the origins of sexual technologies are those that are pushed to the background. Chapter 7, "Legitimizing Sex with Technology: Prisoners, Nazis, Misogynists, and the Origin Stories That Go Untold," argues that one of the primary functions of the tale of the dames de voyage, as it is deployed in contemporary histories, is to legitimize the development and use of twenty-first-century sexual technologies such as sex robots. To illustrate this, I look at a set of alternate stories that these present-day narratives choose to sidestep. The first such story locates the origins of sex dolls in prison, made manifest through examples of actual makeshift sex dolls created by prisoners. The second story, a myth just as false as the story of the sailors' dolls, is the legend of how Nazi leaders supposedly invented the blow-up doll for use by German troops. The third story is the myth of Pygmalion. As mentioned, a close reading of the myth reveals Pygmalion to be not a great classical hero but a ridiculous loner driven by his disdain for women. Together these stories draw to the surface those characteristics of sex tech culture today that other versions of history attempt to ignore: its anxious need to render sexual technologies the domain of "good" subjects and to distance itself from deviance; its ties to the logics of eugenics and white supremacy; and its undercurrents of toxic masculinity that animate much of the techno-utopian rhetoric around high-tech sexual futures.

Despite all of these problems with the tale of the dames de voyage, the story of the sailors' dolls still contains possibilities for a radical reimagining. In the conclusion, "Reclaiming the *Dames de Voyage*: The Feminist Potential of a Fictional Past," I consider how embracing the imaginary nature of this imagined history might open doors to a queerer, more feminist, and more socially just vision of history. To explore this, I offer a reading of an erotic short story by Anaïs Nin, which includes a scene in which a woman fantasizes about being a rubber sex doll passed between sailors.[40] Through this text, we see how the tale of the dames de voyage can be used to flip the script on narratives about technological innovation and their relationships to gender and desire. In this sequence, the woman willfully becomes the doll, blurring her role as sexual subject and sexual object and using her rubber body to both bounce back and "fuck back." In closing, the conclusion reflects on the notion that the origins of sexual technologies lie "at sea." Drawing from writing about the ocean and its intimate relationship to media, I conclude this volume by theorizing what it might mean to set the history of sex tech adrift.

Lessons from the Tale of the *Dames de Voyage*

The lessons that emerge across these chapters extend far beyond the tale of the *dames de voyage*, and indeed beyond any one origin story we might tell about the history of sexual technologies. From this work, we learn that sex dolls have a long history and that, in many ways, it is not the history that has already been told. We also learn that sex tech, as both a set of actual technologies and as a cultural imaginary, has happened before. Previous waves of technological shift have already brought with them visions, simultaneously eager and anxious, of how high-tech devices will imminently change the future of sex. In addition, through this investigation of the dames de voyage, we learn that the history of sexual technologies has never actually been about the inventiveness of straight men. Instead, it is the history of industries that have shaped and been shaped by sexual practices. It is also the history of the sex workers, women, queer people, and people of color whose lives have been bound up with the production and popularization of sexual technologies since long before these technologies came to be seen as the domain of male inventors and consumers. We find as well that sexual technologies have long been built as much on forces of oppression as on self-expression. Of course, those of us who study the cultural implications of contemporary technologies recognize this to be an underlying truth about tech more generally. Yet it is important to recognize that sexual technologies themselves, commonly presumed to stand apart from concerns about the politics of tech, also carry with them a legacy of discrimination that has, over the course of time, come to serve as the foundation for the erotics of technology.

From this investigation, we can also take lessons about how the history of sexual technologies has been made. These lessons likewise apply to the telling of sexual and technological histories more broadly. The way that such histories are told is always inherently political. Feminist historians already know that the way we choose to tell histories of sexuality and gender is inherently subjective, communicating values about whose stories should be told and how. However, the history of technology is just as subjective, even when technology itself seems built on the cold, hard "facts" of machines and computer code. And yet, as the case of the dames de voyage makes clear, even narratives about history that are designed to uphold a discriminatory status quo can be subverted and made radical. Remaking history is a part of the work of reclaiming the future. Indeed, the future of sex tech, like its past, belongs to the very people who have been pushed out of its history. No matter who tries to convince us otherwise (and by *us* I mean my fellow femme and femme-adjacent folks and my fellow queer and trans folks, as well as the people of color and the people living under the ongoing effects of colonialism with whom I stand in solidarity), sexual technologies belong to us today and tomorrow because they have belonged to us for centuries.

I offer these lessons as launching points for anyone interested in the history of sexuality and technology, but more specifically I am speaking to the vibrant, interdisciplinary network of scholars whom this book is both in dialog with and indebted to. In addition to the authors whose versions of sex tech's history I critique, some of my most immediate interlocutors are other feminist historians addressing the material histories of sex toys. Among these are Lynn Comella, Hallie Lieberman, Donna Drucker, Jessica Borge, Anjali Arondekar, Cynde Moya, and, from a former generation of research, Rachel Maines.[41] In addressing sex dolls in particular—and, by extension, sex robots—I have looked to the work of Agnès Giard, Neda Atanasoski, Kalindi Vora, Josef Nguyen, Allison de Fren, and Marquad Smith.[42] There are numerous people who have done important work on gynoid dolls and figures of mechanical women across the history of fiction, art, and film. Among those, I draw primarily from Julie Wosk, Anne Balsamo, Minsoo Kang, and Bliss Cua Lim.[43] Work on all of these topics has served as an invaluable model for how to create a feminist media history that is grounded yet imaginative, authoritative yet self-reflexive, and open to the strange, unexpected joys of sexual and technological histories.

In thinking about technologies like the sex doll that are designed to interface with the body, and specifically to replicate sensations of touch, I draw guidance from work by David Parisi, Rachel Plotnick, and Teddy Pozo.[44] Historians who have tracked shifting relationships between gender and computing, such as Mar Hicks, Janet Abbate, and Ruth Oldenziel, have also helped shape my thinking about how cultural narratives around technology tell different stories about who matters in the realm of tech.[45] In thinking about both the discriminatory and the liberatory potentials of technology, I look to scholars whose work confronts technology from perspectives of race and disability, like Ruha Benjamin, Nettrice R. Gaskins, Safiya Umoja Noble, and Sasha Costanza-Chock.[46] Fellow media historians like Jacqueline Wernimont, Caetlin Benson-Allott, and Carly Kocurek have all offered roadmaps for how to make new meaning from the everyday media practices of the present past.[47] Carlin Wing's research on the history of rubber, both in relation to colonialism and contemporary imaginaries, has deeply influenced my own discussions of the role of rubber in the development of sexual technologies.[48] None of that is to mention the dozens of other scholars who have done and are doing vital research on the infinite number of other ways that technology and sexuality intersect and interweave. They are too many to list here, but their work has invaluably shaped and reshaped my thinking in the many years since I first stepped into the world of sex tech.

Because this book explores what it might mean to locate the origins of sexual technologies at sea, it draws from both historical research in maritime studies and more humanistic work about the ocean. In addition to the many maritime studies scholars whom I reference in chapter 3, work by Sowande' M. Mustakeem on the Middle Passage and Melody Jue on seawater as a medium have proven particularly valuable.[49] Also forming a backdrop to this vision of sex dolls on the open seas is work that has demonstrated the material connection between computational technologies and the ocean itself, such as that by Nicole Starosielski and Alenda Chang.[50] In my thinking about the power of speculation and the possibilities for remaking history, I am inspired by works like Jacob Gaboury's "A Queer History of Computing," which encourages us to envision the queer intimacies of the past without attempting to resolve those visions into historical truth.[51] I am likewise inspired by poetic work like Alexis Pauline Gumbs's writing on how the lives of aquatic mammals jibe with Black feminist thought and Alexis Lothian's discussions of queer speculative futures that remake the future through fantasy.[52] Video game scholars and historians Laine Nooney, Aubrey Anable, and Whit Pow have all helped shape my methodology and my understanding of my own ideological goals in participating in the messy work of making history.[53] In particular, Nooney's and Anable's respective writings on shifting from a model of *excavating* history to a model of *spelunking* have been constant guides for me as I have gone about feeling my own way through remnants of history and creating a picture of the past that is surely imperfect and incomplete but is also true.[54]

A few additional notes about this book: Despite my best efforts, unqualified references to *women* and *men* do make appearances throughout. At times, sex dolls are also referred to as "looking like women" or "being made in the shape of women." As a nonbinary person myself, I am admittedly irked by this description and I recognize its problems—among them, that referring to people and objects in this way perpetuates a vision of sexual history as the history of cisgender people, that it leaves little space for those who fall outside the gender binary, and that it risks reinforcing misguided beliefs that gender can be equated with the shape of one's body or the form of one's genitals. I use terms like *men* and *women* because they are the rhetorical building blocks around which the documents I am analyzing are structured. They are also the terms around which cultural fantasies of the sex doll have formed. I recognize that this is far from a perfect answer, however, and the place of transgender and otherwise gender non-normative people in the history of sexual technologies is a subject that deserves a book all its own. On a more logistical front, it may be helpful to note that many of the sources discussed in this book were originally published in French or German; unless otherwise stated, all translations of these works are my own. Much of the labor of this research has gone into playing academic detective: figuring out strange twists and turns in citationality and making connections across texts where those connections have been intentionally erased. I have tried to hold myself to the same level of citational accountability to which I hold the authors whose writing I analyze. For any readers interested in picking up this work and heading off on the hunt for these sources on their own, I have done my best to leave breadcrumb trails of references in the footnotes.

Like so much feminist scholarship, this work is not only professional but also personal. Searching for and making sense of the dames de voyage has brought together many seemingly disparate elements of my life as a scholar and a writer. My research and training are highly interdisciplinary. In addition to my background in journalism and creative writing, I hold a PhD in comparative literature. I have worked in academic departments of video game design, information science, and film and media studies. I am a queer studies scholar who has had the honor of helping build a flourishing community of queer research around video games and digital media.[55] However, I have also faced discrimination within tech circles for the better part of two decades, not so much for being queer as for being a person assigned female at birth who is far less interested in attaining or celebrating technical knowledge than in asking hard questions about technology and culture. I am a white, able-passing person (with an invisible disability) who recently earned tenure at a major research university. I am the immensely proud mentor of graduate students who are doing the crucial work that will become the future of our fields. I am a person who is deeply skeptical of narratives about sex tech and yet truly gleeful when I find myself immersed in the artifacts of sex tech's real material histories. I am that professor in Southern California who stands at the ready to shower you with endless gratitude and a respectable amount of monetary compensation if you are that one human being who still has an actual inflatable rubber doll from the turn of the twentieth century in their basement. If that is you, I am begging you, call me.

A Final Note on Sex Robots and the Stakes of Reimagining History

It is impossible to talk about sexual technologies or sex dolls today without talking about sex robots. I began this introduction by saying that sex dolls were having a cultural moment. It might be more accurate to say that the cultural moment belongs to sex robots, and sex dolls are along for the ride. Whether we like it or not (I do not), sex robots are everywhere. Well, not real sex robots, at least not for the most part. Rather, it's the idea of sex robots, the fantasy of sex robots, the promise that sex robots are always already just about to be arriving on our doorstep.[56]

"Sex robots are coming." This, often verbatim, is the ominous claim made by a huge array of journalists, authors, roboticists, and advertisers.[57]

Along with this promise come reports of new advances in artificial intelligence for sex robots, innovations in sex robot customizability, and reports that the voluptuous silicone rubber bodies of sex robots now feel and look and even smell more realistic than ever before.[58] Of course, these reports stir up their share of debates. Some argue that sex robots will bring out a "dark side" in their users or that they promote violence against women or children (a case often made by comparing sex robots to prostitutes in ways that disparage sex work) or that they might get hacked and kill the people who are having sex with them.[59] Others counter such arguments by insisting, for example, that robots are good for society because they increase sexual access for people with disabilities or that would-be sex robot lovers are being unfairly judged by small-minded sexual conservatives.[60] Sex robots are also increasingly the stuff of academic research. Representatives from fields like computer science and human-computer interaction are working to develop the technologies that will make the large-scale production of consumer sex robots a viable reality (there are a few early versions already on the market).[61] Meanwhile, scholars in this field are publishing articles and leading conferences that breathlessly celebrate the possibility of "intimate relations with robots."[62]

Let me be honest: I wish that I were not talking to you about sex robots. Although my scholarly perspective is a critical one, I try to work on subjects that I take joy in. That is why I often study video games and internet culture, both of which are close to my own heart. It is also how I have come to write this book; I am deeply invested in exploring how cultures and practices of intimacy shift across media forms, especially ones with ties to digital media. I love learning about unexpected ways that people—either in the present or in the past—have used technologies to connect with one another sexually. I understand and feel compelled by scholarship that argues for the intersection of sex and tech as a space of feminist, queer, and trans possibility. I have immense respect for the women and queer folks and sex workers whose labors and whose longings have driven the development of sexual technologies (and technology more broadly) and yet who are so often written out of technology's history. One of my most prized possessions is my ever-growing collection of antique vibrators. These crumbling technological relics of a sexual past make me happy in ways I cannot quite explain.

What does not make me happy is sex robots. This is not because I object to sex robots or any other given sexual technology on some immutable moral grounds. I am far more sex positive than I am sex negative, if we still believe in such a dichotomy, and I am neither a technodystopian nor a technoutopian. I am not here to dictate whether people should or should not have sex with robots. Nor am I here to pass judgment, as Hillel Schwartz does when writing about sex dolls in *The Culture of the Copy*, on whether or not sex robots should be considered "lovefakes," mere "shoddy forgeries" inadequate to the task of replicating the complex realities of sex.[63] The problem with sex robots, in my opinion, is not sex robots. The problem with sex robots is people.

In her essay "Your Robot Isn't Neutral," Safiya Umoja Noble explains how the design of robots today is driven by stereotypes. Both in their physical appearance and in their data-driven, algorithmic artificial intelligence, says Noble, robots are being developed around cultural norms that reinscribe troublesome hierarchies of race and gender at the levels of both software and hardware. Unfortunately, despite the crucial relevance of these issues to robotics research, "the field of robotics rarely engages with social science and humanities on gender and race." This shortsighted rejection of socially engaged perspectives on the part of computer scientists is resulting in the unchecked spread of racism, sexism, homophobia, ableism, and economic exploitation in the design of robots, even those developed for "social good." These issues are inseparable from the sexualization of robots. "It's no surprise that we see a host of emergent robotic designs that are pointing toward women's labor," writes Noble, including "doing the work of being sexy and having sex." After all, Noble reminds us, "robots are the dreams of their designers, catering to the imaginaries we hold about who should do what in our societies."[64]

These same concerns about robots more broadly are amplified when it comes to sex robots in particular. Contemporary discourses about sex robots are fundamentally bound up with cultural politics of gender, sexual identity, and race. Scratch the surface and you will find that what most people are talking about when they talk about sex robots is not sex robots at all. Proponents of the so-called rise of sex robots, and there are many of them, paint a picture of a bold new sexual future sitting just over the horizon, one in which both the sex we have and the sexual partners we have it with finally fulfill our wildest dreams. Yet these "wildest dreams" are of a very particular sort. Invariably, they are the dreams of straight, white, cisgender men longing for a sexual partner who is fully customizable, who has no pesky desires of her own, and who participates in what Josef Nguyen describes as "the fiction of a subject capable of consent."[65] The sex robots being developed today may be technologically innovative, for whatever that is worth, but they are socially regressive and reactionary. They reify rather than challenge cultural norms related to identity and power. They leave no space for those who desire to form a sexual connection with something other than a supposedly realistic human-like object whose design is based on extremely straight, extremely white, and largely ridiculous notions of female beauty.

As an academic, the thing that riles me up the most is the growing ties between these questionable attitudes toward sex robots and areas of academia associated with technology. Some of the very same people who have been most influential in promoting sexual robotics research and the need for scholarly investigations of love and sex with robots have participated in misogynistic online harassment campaigns, such as those against women working in video games.[66] They have also been involved in incidents of intimidation in which women scholars were targeted and harassed for raising extremely valid questions about the ethics of sex robot–related research. Every time I see yet another publication written by sex robot evangelists or apologists from STEM fields (fields in which top publication venues theoretically pride themselves on rigorous reviews of objectivity and data), I find myself torn between laughing and screaming.[67] These are fields that are already extremely male-dominated, and topics like feminism, queerness, and racial inequality are still largely taboo there. It's inappropriate to talk about gender issues in computer programming languages, but we can talk about sex robots?[68] That, right there, is the problem, not whether or not any given human has sex with a machine.

This is why I hate talking about sex robots, and this is why we need to talk about them. By changing the ways that we think about the history of sex tech, we can intervene in a present-day echo chamber of fantasies and anxieties that reaches a fever pitch around the figure of the sex robot. I argue throughout this book for the importance of attending to imaginaries: the cultural imaginaries that have surrounded various generations of sexual technologies, the origin stories that turn out to be imaginary, the power to remake history by imagining it otherwise. This issue of imaginaries is also central to understanding the sex robot and to challenging its ascendancy. The cultural life of sex robots, today as across history, has always been far more about how sex robots are imagined than how sex robots really are. It is imaginaries like these that require our careful interrogation and insistent dismantling.

And, luckily, it is here that history itself becomes most useful. The history of sex robots is also the history of sex dolls, and the history of sex dolls reveals many things. It reveals that today's promises that sexual technologies will be more pleasurable than ever are the same ones that have been made for the last century and a half. It reveals that romanticized visions of sex robots and other devices like them have always been tied to a precarious straight, white masculinity that has attempted (and failed) to keep femininity, queerness, and racialization at bay. It reveals just how much work it is has taken, over the course of decades, to make sex robots seem "natural" and good. If we see the sex robot as the epitome of sex tech, and if we see sex dolls as the sex robot's predecessor, then the origin stories discussed here become stories about the birth of sex robots. By setting the terms of the sex robot's cultural birth, they attempt to set the terms for the sex robot's ongoing cultural life. In disrupting established stories about the sex robot's birth, then, we take an important step toward bringing about the sex robot's death: the death of a certain set of narratives that position the sex robot, of all possible technologies for experiencing and expressing sexuality, as the gold standard of sex tech. Death to the sex robot. Long live the sex tech we have not yet had the freedom to imagine.

I Searching for the Very First Sex Doll

1 Contemporary Tales of the *Dames de Voyage*: The History of an Imagined History

A ship sails on the high seas. Its crewmembers, a group of men bunked together in tight quarters, spend their nights longing for the embrace of the wives and sweethearts they left behind when they undertook this long journey. Maybe these sailors are on a Dutch ship, or maybe it's a Spanish one, or maybe it's somehow simply European. Maybe this ship is part of a fleet in the midst of fighting a war, or maybe it's on its way to Asia, traveling along a colonial trade route. This might be the 1600s or the 1700s or perhaps the 1800s. Whatever the case, the scene remains the same: lonely and far from home, these sailors use their ingenuity and the limited resources at hand to make an alternative to a flesh-and-blood woman. They build a rudimentary life-sized doll, constructed from cloth, or maybe leather, and possibly stitched together with twine, with a hole cut between her legs. Passed back and forth between sailors, she becomes a communal lover. Over time, her innards (loose cotton or straw stuffing?) collect dried bodily fluids and begin to fester. What happens to her once she has become too tattered to use or once the crew completes their journey is unknown. Yet her place within the history of sexual technologies remains fixed: she is the "very first sex doll," the origin point from which all sex tech supposedly follows.

This is the tale of the *dames de voyage* as it appears in a striking number of contemporary accounts. Or, more accurately, this is a composite version of that tale, since the supposed facts vary from telling to telling. In the introduction, I explained that the story of the sailors' dolls has come to hold a privileged place within narratives about the history of sex tech. Here, I delve into the tale itself—a first step in the search for the dames de voyage—by mapping its manifestations across a network of twenty-first-century texts. These include popular histories, news and online articles, and academic writing published roughly between 2006 and 2021 by authors in North America and Western Europe. On the surface, the story of the very first sex doll appears to be a simple one. Indeed, each time it is repeated, it is typically presented as a brief historical anecdote—a quirky story that grounds sexual technologies in the past. However, tellingly, the story of these supposedly lustful, inventive men at sea has rarely been told the same way twice. As I demonstrate in the first half of this chapter, comparing these texts makes clear that there are, in fact, numerous versions of the story in circulation, many of which contradict each other. These contradictions reveal the cracks in the tale, raising questions about its veracity but also drawing attention to its constructedness: the different ways the story is told, the varying contexts in which it appears, and the ends to which it is deployed.

The second half of this chapter turns from the question of *what* (i.e., what is the tale of the dames de voyage as it appears in this network of texts?) to the question of *how*: How did this tale form, and what can we learn by parsing out its evolution? To answer these questions, I work backward, starting from the most recent writing to recount the story of the sailors' dolls and moving through a murky, tortuous trail of citations and influence. If such a task appears, at first glance, like a standard part of the research process, the reality is far more complicated. In doing this work, I have often felt like an obsessive investigator in a procedural drama connecting dots on a corkboard with string. As I demonstrate ahead, mapping the citational trail that brings us the contemporary tale of the dames de voyage has required considerable sleuthing precisely because, as it turns out, this trail is characterized by slippages, omissions, and intentional obfuscations. What emerges from this process is a history of an imagined history: a revealing picture of how an origin story forms not so much around facts as around bids for legitimacy and originality and the gender politics of citation.

Ultimately, the work presented in this chapter serves as a first step toward locating the dames de voyage and their actual place within the history of sexual technologies. Yet in truth, it tells us more about the present than about the past. Most of all, we find a compelling reminder that the making of history is a living, shifting, messy thing. This is especially true for histories of sexuality and histories of technology, both of which are often closely tied to everyday cultural practices that are rarely archived and whose historical afterlife remains precarious. The story of the sailors' dolls becomes a case study in miniature for understanding how such histories circulate—and how paradoxically, through their very circulation, they come into being. Rather than trying to reconcile these various accounts, my goal in this chapter is to demonstrate how the making of sex tech's history has operated like a game of telephone: an appropriately intimate, mediated simile for the telling of a story that supposedly predates contemporary media cultures and yet is deeply intertwined with media. Like children sitting in a line, heatedly passing a whispered message from ear to ear until the garbled result bears only a distant resemblance to the original utterance, the texts discussed here create a chain, changing the tale of the very first sex doll as they move it down the line, taking creative liberties in its telling and making it into something new.

Telling the Tale of the *Dames de Voyage*

The tale of the dames de voyage makes numerous appearances across many twenty-first-century texts, manifesting each time in different ways and situated within different contexts. Some of the most prominent examples include popular histories of sex and sex tech—books written for a mainstream audience. Notable among these is David Levy's 2007 *Love and Sex with Robots*.[1] *Love and Sex with Robots* is a book on a mission: to convince readers that sexual and romantic relationships between humans and machines will soon become commonplace, socially acceptable, and far more thrilling than sex between human partners. By roughly 2050, Levy insists, "Love with robots will be as normal as love with other humans . . . [and] robots [will] teach more than is in all of the world's published sex manuals combined."[2] In an attempt to demonstrate the long-standing precedent for humans engaging in sex with machines, Levy gives his version of the history of sexual technologies, focusing on the advent and popularization of various sex toys. He opens with a discussion of early vibrators, which, drawing from Rachel P. Maines's *The Technology of Orgasm*, he describes as emerging from a nineteenth-century history of medicalizing women's pleasures.[3] However, in an attempt to complicate Maines's account and shift focus toward sexual technologies for men, Levy asserts that equal consideration should be given to "sex toys for the boys," such as artificial vaginas and *fornicatory dolls*, a common alternate term for sex dolls found predominantly in sexological writing.[4]

Levy makes references to the dames de voyage at multiple key moments in his telling of this history. Although he briefly describes earlier examples of "women substitutes" being created in Japan, Levy suggests that the Western history of the sex doll traces back to none other than the dames de voyage, made "in the female form" and used by European sailors.[5] He describes the dolls and their usage in this way:

> Employing fornicatory dolls while traveling became popular in Europe during the late nineteenth century, particularly among sailors. The sexual life of sailors has never been an easy one, living and working as they do in an entirely male environment, their trips ashore to the red-light districts of various ports providing just about their only female sexual comfort. Wives and lovers at home could only rarely be visited, so long were the voyages to and from the ships' destinations on other continents. Hence the need for *dames de voyage* (traveling women) as the French called them. . . . These were dolls in the female form, most often made of cloth and used as sexual outlets for sailors on board ship.[6]

This story, as low-tech as it sounds, represents an important step along Levy's teleological timeline of sex tech. It strategically positions the dames de voyage as a pivot point for moving into claims about how sexual devices later became feats of technological prowess. Directly after describing the dames de voyage, Levy writes that "sex dolls of a less primitive form gained a certain measure of popularity in late-nineteenth-century France": supposed wonders of mechanized design with complicated inner workings that purportedly simulated the experience of vaginal intercourse.[7] These too, in Levy's account and others, get grouped under the heading *dames de voyage*. By placing his description of the sailors' dolls within a larger narrative about invention and progress, Levy interpolates the dolls into a genealogy that leads up to the "high-tech" sexual devices of the present day. At the same time, Levy's description of the dames de voyage imbues them with a sense of supposed timelessness—providing a precedent for the idea that (straight) men have always desired (cisgender) women and are therefore naturally inclined toward sex with machines.

Of particular note in *Love and Sex with Robots* is Levy's inclusion of an image that appears to represent the sailors' sex dolls themselves (see figure 0.2).[8] A grainy, black-and-white reproduction of what seems to be a series of three photographs, it looks to depict three such dolls dressed in women's clothing, one with a hole cut at the pubic area of her long white bloomers, presumably to allow for insertion. On the image, there is a handwritten annotation in flowery German script, as well as what appears to be an earlier printed description of the image, also written in German. These elements of the text are nearly illegible, however. Levy has labeled the image simply "fornicatory dolls." However, his placement of the figure alongside his discussion of the dames de voyage clearly suggests that the image is meant to represent the specific dolls he is describing, either in their form as sailors or as later mechanized creations. This image, which I will return to at a number of later points, has become a prominent feature in this broader network of contemporary texts. Authors frequently replicate and reprint it, often explicitly labeling it as *dames de voyage*, making it (what is presumed to be) the one material trace of the dolls' actual existence and a visual stand-in for the dolls themselves in their contemporary cultural life.[9]

Another influential popular history that features the tale of the dames de voyage is Anthony Ferguson's 2010 book *The Sex Doll: A History*.[10] Whereas Levy addresses sexual technologies more broadly, Ferguson focuses on sex dolls. Ferguson too positions the dames de voyage as an origin point in an imagined historical lineage leading up to present-day technologies, one that supposedly stretches from the sex doll's conception at sea "to the robotic effigies which are now becoming more readily available via the Internet."[11] Ferguson opens the preface to his book by writing:

> It may come as a surprise that there is very little serious literature available on the origins of sex dolls, yet the idea of creating an ideal love object has recurred frequently in writings since ancient times. . . . This is understandable. From the germ of male desire sprang the urge to create something which was female in appearance, but completely receptive and non-judgmental. Yet . . . little has been written about sex dolls from their origins as cloth or sack effigies (*dames de voyage*) utilized by randy sailors on lengthy and lonely sea journeys to their sudden reappearance on the periphery of popular culture around the 1970s as inflatable dolls.[12]

Similarly situating the dames de voyage within a tradition of naturalized "male desire" that purportedly stretches back to "ancient times," Ferguson positions the sailors' dolls as proof of the long history of sex toys that replicate women's bodies for the pleasure of men. Ferguson also deploys his reference to the dolls as proof of the validity of his subject, reflecting an impulse to legitimize both sex dolls and research on the topic. Ferguson's description of the dames de voyage as "completely receptive and non-judgmental" speaks to a fragile masculinity that underlies even this supposed celebration of the sex doll. In this telling, the dames de voyage

become not only the first sex dolls but also the first lovers who could neither turn down a man's advances nor laugh at his desires—a representation
of history that itself worrisomely overlooks how racialized and gendered
subjugations have long forced victims of sexual assault to present themselves as "receptive."

Soon after, in his first chapter, titled "Origin of the Species," Ferguson
dedicates an entire subsection to the dames de voyage. The passage bears
quoting at length as its details will become important for tracing connections to other texts:

> The modern sex doll has its most direct antecedent in the cloth fornicatory dolls
> used by sailors on long voyages. The *dame de voyage* or *dama de viaje* was originated
> by French and Spanish sailors at the height of their respective naval empires in
> the seventeenth century. This was the beginning of the modern sex doll. These
> elementary sex dolls were made from cloth or old clothes and would have been
> quite rudimentary. Imagination would have been critical to their use, particularly
> for a man isolated at sea on a long voyage. Women were generally considered
> bad luck on ships, and every voyage was lengthy. It was very convenient to store
> something as light and malleable as a life-sized cloth doll in the cramped quarters
> of a ship. It is likely that a single doll might have been shared among several
> men, although it is uncertain whether the doll(s) would have been made available
> to the common sailors or restricted to the higher-ranking ship's officials. What-
> ever the case, the dolls could not have been particularly sanitary after multiple
> uses. . . . Like the inflatable doll, the *dame de voyage* would have required of its user
> a suspension of disbelief. Yet this is plausible given the level of sexual frustration
> which would have built up during the lengthy sea voyages of the pre-Victorian
> world. When the only alternatives were masturbation or buggery, perhaps the
> *dame de voyage* presented a viable option to the sexually deprived. In this regard
> we can comprehend why men dreamed up stories of sirens and mermaids to fuel
> their fantasies on long lonely nights in their bunks.[13]

The picture of the dames de voyage that Ferguson provides is rich with
historical "facts," many of them alluringly tangible. According to Ferguson, the dames de voyage were often made from old clothes; they were
lightweight, bendable, and easy to store in tight living quarters. Yet what
is most striking about this passage is how it draws out the role of fantasy
in animating the tale of the dames de voyage. Ferguson writes that using
the dolls would have required a "suspension of disbelief" and likens them
to mythical sirens and mermaids. However, Ferguson too is engaging in a
kind of erotic daydream, imagining whether the dolls were shared between
soldiers, who used them, how dirty they were, and how sex with the dames

de voyage was surely used to fend off the looming threat (or, perhaps, temptation) of "buggery."

In addition to Levy's and Ferguson's works, more recent books, like Hallie Lieberman's *Buzz: A Stimulating History of the Sex Toy*, also recount the tale of the dames de voyage.[14] Lieberman's own focus is on issues of production and commerce surrounding sex toys in the United States. Yet Lieberman too presents a similar narrative about the development of sexual technologies across history, beginning with the supposed sexual habits of cavemen and moving up to the high-tech sex toys of the twenty-first century.[15] In *Buzz*, the dames de voyage appear yet again as part of a lineage of "toys for boys." Writes Lieberman, "By the 19th century, cloth sex dolls, known as *dames de voyage*, were carried aboard European ships so the sailors could have sex with them."[16] Although this reference is made only in passing, its context is important. Immediately prior, Lieberman tells the story of a blow-up-doll salesman who began selling the dolls for "all the lonely guys who weren't getting laid" during the sexual revolution of the late 1960s and early 1970s. "He thought there were enough profits from the dildo business," Lieberman explains, and "it was time for men to get their chance" (figure 1.1).[17] This juxtaposition suggests that we might see the dames de voyage too as part of a history of men trying to "get their chance," calling to mind the misogynistic rhetorics of twenty-first-century groups like men's rights activists, pickup artists, and self-described "incels," who often justify harassment and assault on the basis that women owe men sexual satisfaction.[18]

In addition, the tale of the dames de voyage crops up in popular histories written for non-English-language audiences. Like Ferguson's *Sex Doll*, Julien Arbois's 2016 book *Dans le lit de nos ancêtres: Sexualité, moeurs, et vie intime d'autrefois* (In the bed of our ancestors: Sexuality, customs, and the intimate life of times gone by) includes an entire chapter subsection dedicated to the dames de voyage.[19] Arbois's book presents itself as an entertaining history of sex, broadly speaking, rather than a history of sex toys or sexual technologies—akin in spirit to Kate Lister's *A Curious History of Sex*—yet its account of the sailors' dolls sounds quite familiar:[20]

> The history of the first sexual dolls in Europe began in the seventeenth century, when Portuguese and Dutch explorers discovered the charms of Asian life and became interested in certain objects in particular . . . Dutch sailors seem to have been the most interested in these practices. It is well known that sailing with

Figure 1.1
An advertisement for an inflatable sex doll sold through a California company. This is a doll of the sort produced so that, as Hallie Lieberman writes, men could "get their chance." *Source:* Paul Tabori, *The Humor and Technology of Sex* (New York: Julian Press, 1969).

women on a ship is bad luck, and these sex dolls could be of service. From then on, [sailors] made substitutes of their own, amalgams of cloth and leather that they nicknamed the *dames de voyage*. These objects, easily transportable and shared between sailors, were found in the holds of ships that crossed the oceans, and in the majority of European fleets, notably the English fleet.[21]

Arbois's claim that the creation of the sailors' dolls was inspired by the "charms of Asian life" is one among many instances in which Asia (often especially Japan) makes appearances in these histories. Already in this passage, we can see the problematic power dynamics of this imagined relationship between European sailors and Asian cultures taking form. Arbois describes the Dutch colonialists as "discovering" their inspiration for the dolls in Asia, and then credits them with the ingenuity to create the dames de voyage—thus positioning European sailors, rather than Asian people, as the early innovators responsible for the origins of sex tech.[22]

Increasingly, texts recounting the tale of the dames de voyage have taken the form of news articles and blog posts, which circulate more freely and rapidly online. There are many examples of these texts, and together they form something of an echo chamber: linking to one another and largely repeating similar versions of the story. Some do, however, tweak the tale in retelling it. For example, in a 2017 blog post titled "Petite histoire de la poupée érotique" (A short history of the erotic doll), self-described amateur historian Priscille Lamure writes that Dutch sailors came up with the idea to make dolls shaped like women from "old pieces of cloth," which they did to beat the boredom of long maritime crossings.[23] "Called *dames de voyage*," says Lamure, "these sex dolls were taken aboard ships and were subjected to multiple assaults from sailors. They quickly became, as is inevitable, nests of germs. Nevertheless, this practice spread to most European fleets at the time."[24] Lamure points to Arbois as the source of her information about the dames de voyage, yet she also puts her own spin on the tale, as evidenced by her note that dolls were "subjected to multiple assaults"—quite a contrast to the romanticization (and fetishization) of the dames de voyage seen in texts like Levy's and Ferguson's.

Another article, Julie Bech's "A (Straight, Male) History of Sex Dolls," published in the *Atlantic*, makes its feminist critique of the dames de voyage even more explicit.[25] Much of Bech's article focuses on Real Dolls, the contemporary brand of high-end sex doll, but Bech also situates these dolls within a longer lineage. She writes, "Throughout history, men . . . with an inclination to make love to women-shaped things . . . have made do in various ways. Sailors often used cloth to fashion fornicatory dolls known as *dame de voyage* in French, or *dama de viaje* in Spanish. . . . Though sailors' dolls were just generic substitutes for the female form—any female form."[26] Like Lamure's "Petite histoire," Bech's account passes judgment on the creation and use of the sailors' dolls, making them the object of biting remarks about men's desire to "make love to women-shaped things." In keeping with this spirit, the introductory text for Beck's article as it was published online reads: "Since ancient times, men have been getting it on with synthetic women. Is this just fancy masturbation, or something more troubling?"[27]

However, even as such articles question the gender politics of sexual histories, they rarely question these histories themselves. It is telling, for example, that nearly identical versions of the tale have appeared in decidedly less feminist contexts. For instance, in 2016, *Penthouse* ran its own article about the history of sex dolls—fittingly as part of a "tech issue," featuring on its cover a nude woman holding, of all things, a retro *Donkey Kong* Game & Watch console over her pubic area.[28] Couched with a discussion about how Nazis supposedly invented the blow-up doll (an alternate origin story that will return in chapter 7), the article states: "There have been men throughout the ages who have used woman-shaped objects to sate their sexual desires. . . . For example, sailors on long trips would fashion love puppets out of cloth. These 'companions' were amorously known as *dames de voyage* in French or *damas de viaje* in Spanish—journey women."[29] In addition to this slightly misleading translation of the French and Spanish terms as "journey women," the *Penthouse* article illustrates how strikingly similar versions of the tales of the dames de voyage are presented in strikingly different publications.

These popular histories and mainstream articles have brought the story of the sailors' dolls to scholarly texts as well, where it is being codified through academic publication. Consider the 2017 collection *Robot Sex: Social and Ethical Implications*, a volume that opens with a simple yet ominous assertion: "Sexbots are coming."[30] The first essay in the collection, written by coeditor John Danaher, is titled "Should We Be Thinking about Robot Sex?"[31] In this article, yet again, questions of present-day and future sexual technologies are framed through a recourse to the past. Danaher writes: "Artifacts for sexual stimulation have long been a staple in human life. . . . In 1869, the American physician George Taylor invented the first

steam-powered vibrator. . . . At around the same time, the first manufactured sex dolls became available, though the idea of the sex doll has a much
longer history—one that can be traced back to the myth of Pygmalion
and to Dutch sailors' *dames de voyage* in the 1700s."[32] As in other texts,
the dames de voyage function in Danaher's article as a way to historicize
sex tech while also shifting away from other versions of history, sidelining devices made for women and centering devices made for men. In such
moments, we see how the tale of the dames de voyage is put to work not
only in the service of "making technology masculine," to use Ruth Oldenziel's turn of phrase, but to make the history of technology masculine as
well, a phenomenon explored at greater length in chapter 5.[33]

A Tangle of Contradictions

Reading examples of these accounts side by side is a curious experience.
They seem, at first, so similar—like mere repetitions. Indeed, there are many
patterns that reappear across these texts, and they share many related concerns. For example, all of these accounts are, in their own ways, invested
in the performance of historicity: the presentation of specific, seemingly
concrete facts that create a sense of historical realness for the history of
sex tech. Yet if we look closer, the dames de voyage as they appear in such
works quickly become a tangle of contradiction. To begin making sense of
this cacophonous array, I linger with this network of texts to draw out their
shared concerns and tensions. Holding them side by side makes visible
their slippages. It also highlights ideologies that shape these stories. While I
engage with these ideological implications in more depth in later chapters,
it is useful to begin unpacking such issues here, while the details of these
contemporary accounts are still fresh in mind.

Place, for instance, is a recurring issue in these stories about the sailors'
dolls. In all of these tellings, the dames de voyage are linked to Europe.
However, which European locales the dolls are associated with differ from
text to text. Some authors (like Levy and Lieberman) simply say *Europe*,
though the use of the French phrase *dames de voyage* implies a particular
connection with France.[34] Other authors (like Ferguson and Beck) offer a
Spanish version of the phrase as well, suggesting that the dolls were also
used by Spanish-speaking sailors.[35] Arbois names Portuguese and Dutch
"explorers" as the first to begin using sex dolls on ships but also says that

the English had a notable proclivity for the dames de voyage and that they could, eventually, be found in the majority of European fleets, a remark found in Lamure's "Petite histoire" as well.[36] Which is it? Were these sailors' dolls the inventions of the French, the Spanish, the Portuguese, the Dutch, the English, or simply Europeans?

This proliferation of possible national origins suggests that the question of *where* is, in reality, less a matter of historical fact and more a matter of acceptable otherness. That is, the use of this oddly floating signifier, which positions the sailors' dolls among some European population or another, allows those who tell and retell the tale to situate the dolls in a place envisioned as geographically distant enough to be othered (don't worry, sailors' sex dolls weren't invented by us!) and yet close enough for these authors to claim cultural affiliation (but they were invented by people like us). It is revealing just how infrequently the authors who tell this tale cite their *own* nations as the ones that invented the dame de voyage. To Americans and English-speaking Europeans, the dolls are the domain of the French and the Spanish. To the French, they are the domain of the Portuguese, the Dutch, and the English.

References to time periods represent another touchstone across these accounts. All of the texts described cite some specific time period when sailors supposedly made and used these dolls. However, few accounts agree on when that time was. Levy says the dolls were a product of the late nineteenth century.[37] Lieberman states simply the nineteenth.[38] For his part, Ferguson jumps back two to three hundred years, writing that the practice of using dames de voyage was at its peak in the seventeenth century, at what he refers to as the height of the French and Spanish naval empires.[39] Arbois states that these practices *started* in the seventeenth century.[40] And Beck asserts that the Dutch were using sailors' dolls in the seventeenth century but that they were also used in Europe more broadly during the general "past."[41] This sense that the dames de voyage are simply the stuff of some vague time long ago also manifests as hand-waving gestures of verb tense and rhetoric. The author of the *Penthouse* articles writes merely that using sex dolls is something that sailors "would do," while Lamure's short history describes the creation of the dames de voyage as something that happened "at that time."[42]

As with the question of geography, a survey of these texts reveals no clear consensus about when the dames de voyage are supposed to have existed. This scattershot array of inconsistent facts reflects an investment more in cultivating a feeling of historical authenticity (made "real" by the use of dates) than in history itself. At the same time, the instrumentalization of temporality in these accounts serves a curiously contradictory set of purposes, as we see in Levy's text. References to time are meant to make the dames de voyage—and, by extension, sex toys and sex tech—seem both credibly old *and* functionally ageless. Such histories suggest that contemporary sexual technologies have a history so long that it serves as evidence that things have always been this way. In this sense, the tale of the dames de voyage is simultaneously an origin story and a story used to argue that sexual technologies need no origin.

Also recurring in these accounts are details about the physical materials out of which the dames de voyage were reportedly fashioned. Levy and Lieberman both describe the dolls as being made of cloth.[43] Ferguson gives more detail, stating that the dolls were made of "cloth or old clothes" or "sacks."[44] Arbois claims that the dolls were made of both cloth and leather.[45] Beck writes only that they were made of leather, describing them as "leather masturbation puppets."[46] Conjecture about what materials were used to fill the dolls also pops up across these texts; some say straw, while others say cotton stuffing. These details imbue various versions of the tale with different cultural and affective qualities. Within the Western context of the tale's production and reception, cloth is predominantly feminine-coded, associated with traditions of sewing, crafting, and "women's work."[47] It suggests that the dolls were soft, flexible, and yielding. By contrast, leather has been traditionally coded as masculine, representing a kind of "hard" masculinity, though leather's association with certain gay male subcultures complicates any notion of leather as simply heteronormatively masculine.[48] The idea that the dames de voyage were filled with cotton stuffing links them to the *dakimakura* of today: life-sized body pillows, often with images of anime-style women characters, for which one can purchase inflatable inserts complete with vaginal attachments that transform the pillows into toys for penetrative sex (figure 1.2). By contrast, embracing a doll stuffed with straw would be a pricklier, itchier business.

These material details serve the purpose of making the dolls seem real (physical, tangible, an earlier version of the "realness" that we find in the hype around today's increasingly "realistic" sex robots) and making the sailors seem resourceful: masculine protohackers and precontemporary

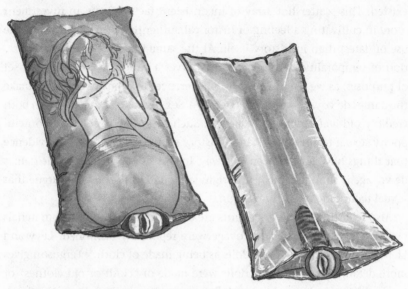

Inflatable insert with replica vagina for body pillow

Figure 1.2

Artist's rendering of a *dakimakura* body pillow with inflatable insert including replica vagina for penetration. *Source:* Original art by Kathryn Brewster.

"makers" capable of cobbling together a sex doll simply from the odds and ends found on a ship.[49] At the same time, details about materials perform a kind of low-tech authenticity, suggesting that the origins of today's sexual technologies can be found not in the realm of the computer but rather in crafty, salt-of-the-earth constructions of threadbare cloth and briny scraps of leather. And yet in some of these tales, such attempts to use references to materials to root sexual technologies in an imagined butch, predigital era end up backfiring. Ferguson, for example, claims that the dames de voyage were made out of old clothes. Whose clothes? Are we to envision that these dolls shaped like "women" were constructed from the discarded outfits of male sailors? This too suggests a far less straight vision of the "very first sex doll": a penetrable partner made from men's clothes, worn thin from time spent pressed up against men's bodies (figure 1.3).

Figure 1.3
A caricature from roughly 1800 of a British sailor in 1783, demonstrating the sort of clothing some contemporary accounts suggest were used to construct makeshift sex dolls.

In addition to sharing an interest in specific material descriptors, these various accounts reflect a common investment in certain ideologies. For instance, despite the fact that some accounts of the dames de voyage are voiced by feminist writers or scholars, sexist logics and misogynistic rhetorics continue to find their way into this network of texts.[50] These logics manifest, at times, as assertions that technological innovation is driven by men's natural desire to use women as objects. At other times, it is reflected in comments about sexually transmitted infections. Levy, for instance, writes that sexual "substitutes" are helpful tools, not just in some historical past but in the present moment, for men who are away from their wives because they can "[satisfy] one's sexual needs without creating any cause for concern about disease."[51] Here, sex with dolls and robots is presented as logical and responsible, whereas sex with flesh-and-blood women is associated with sickness—a not-so-subtle jab at sex workers, who are often presented as a foil to the dames de voyage.

We can also see across these texts that such accounts are commonly invested in rendering the history of sexual technologies straight: the stuff of men doing manly things (sailing the high seas, feeling lustful, inventing gadgets) while desiring women. That drive toward heteronormativity likewise encompasses a hefty dose of homophobia. Most of these texts described here conspicuously ignore the homoerotics of the imagined ship environment they describe—or they dismiss sex between men as so unappealing that sailors surely would have preferred to sleep with dolls.[52] In such accounts, the possibility of sex between men is the specter that looms over the tale of the dames de voyage and, by extension, the larger imagined history of sexual technologies. Each time the story of the sailors' dolls is retold, regardless of its geographic or temporal details, it contributes to the insistent, anxious work of situating sex tech's origins in opposition to queer sex.

Legitimizing Writing on Sex Dolls: Slippages and Obfuscations in the Citational Lineage

The tale of the dames de voyage is both pervasive and oddly inconsistent. How do we make sense of these contradictions? Which version of the story of the sailors' dolls (if any) is true? What are the origins of this origin story? One way to answer these questions is to embark on a journey back from the present to trace the lineage of citations and textual influences that has brought this contemporary tale into being. Tracking the path of the tale's transformation, it turns out, is far from easy or straightforward. The intellectual genealogy through which the tale has formed is twisting and muddled, designed to send investigators like myself on wild-goose chases or toward dead ends. Yet it is crucial to understand these slippages and absences because they have come to fundamentally shape the tale of the dames de voyage today. This demonstrates how accounts of sexual histories, as well as histories more generally, are formed around and contorted by the motives of their authors—motives that are themselves tied to cultural forces associated with issues of sexuality and gender.

Let's begin with what we might regard as the most recent microgeneration of texts to recount the story of the sailors' dolls, those published roughly between 2014 and 2021. Some of the texts that fall into this category include Danaher's "Should We Be Thinking about Robot Sex?," Arbois's *Dans le lit de nos ancêtres*, and articles from the *Atlantic* and *Penthouse*.[53] Tracing the citational trail from these most recent texts reveals that a sizeable percentage draw their information from a shared source: Ferguson's *The Sex Doll: A History*, published in 2010.[54] Some of these texts do cite Ferguson directly, or cite authors who cite him. However, others clearly draw from Ferguson's writing but do not cite him.

Making matters murkier, even when authors do cite Ferguson for their claims about the dames de voyage, it is not uncommon to find that their facts do not actually match those found in Ferguson's book. Consider the case of Danaher's essay "Should We Be Thinking about Robot Sex?" To support his claim that sex robots have a long history "that can be traced back to . . . Dutch sailors' *dames de voyage* in the 1700s," Danaher includes an endnote pointing to Ferguson's *Sex Doll*.[55] Yet if we hold Danaher's account against Ferguson's, the two do not line up. Ferguson says the sailors' dolls were made by the French and Spanish in the seventeenth century, whereas Danaher says that they were made by the Dutch in the eighteenth century. From where then is Danaher drawing his version of the tale? And why would he cite Ferguson when he is pulling from elsewhere?

The answer has two parts—one technical, one cultural. First, if we connect a few dots, we find that Danaher was likely not drawing from Ferguson at all, but rather from Bech's "A (Straight, Male) History of Sex Dolls" from 2014. Bech, for her part, first mentions the dames de voyage and then later mentions other sex dolls supposedly used by Dutch sailors. It appears that Danaher has conflated the two elements of Bech's article, collapsing them into a shorthanded description of the dames de voyage as sailors' sex dolls used by the Dutch sailors. For her part, Bech does mention Ferguson, suggesting that she has drawn her historical information from *The Sex Doll*, though she herself confuses different elements from Ferguson's text.[56] This explains how Danaher's essay ends up with its citation to Ferguson: Danaher pulled information from Bech's article and then cited the source that she cited without checking that they matched.

This may seem like a small, pedantic mix-up to belabor, but it's representative of a larger trend. The citational lineage of the tale of the dames de voyage is in fact chock-full of such slipups and slippages: facts spliced together, references that go nowhere, information given without evidence. These are not just marginal features of this citational trail; they fundamentally characterize how the tale of the dames de voyage has moved between contemporary texts. Such messiness reflects, in part, a larger laissez-faire attitude toward the telling of sexual histories, wherein facts are implicitly understood as always up for alteration and embellishment and citations are wobbly at best. (Certainly, contemporary scholars are not the only ones guilty of this; we might equally point the finger for this offense at Michel Foucault's *History of Sexuality*, for example.)[57] When it comes to writing the history of sex, the case of the dames de voyage suggests that writers often operate around the uninterrogated presumption that they can tweak the details of sexual anecdotes without particular concern for accuracy because sex itself is viewed as a kind of universal truth. The effects of this attitude are all the more difficult to correct when the sexual histories in question are seen as particularly idiosyncratic or niche, since present-day readers are unlikely to scrutinize their veracity.

Danaher's choice to cite Ferguson rather than Bech also exemplifies another feature of this citational lineage: obfuscations. We might guess that a scholar who gives an incorrect citation is doing so in order to throw readers off their trail—inserting a false source so they can present information with impunity. This is not the case here, however. Although Danaher's description of the dames de voyage is questionable, it is not made up. Instead, what we can find reflected in this choice is a bid for legitimacy— which becomes all the more pressing for an author or scholar writing about "illegitimate" topics like sex tech, sex dolls, or sex robots. Legitimacy, in the context of scholarship and historical writing, is accrued and performed in many ways. Mapping the citational trail behind the tale of the dames de voyage highlights the close ties between notions of legitimacy and citationality. Scholarly legitimacy manifests as decisions about which works to cite (e.g., academic vs. popular texts), how to render scholarship itself legitimate (e.g., through providing citational notes for statements that might otherwise be viewed as questionable), and whether to lend legitimacy to authors whose perspectives differ from one's own by citing them. Alternatively, one might implicitly argue against the legitimacy of an author by drawing from their work while declining to cite them—a practice rendered explicitly feminist in writing by scholars like Sara Ahmed and Amanda Phillips.[58]

In crediting Ferguson rather than Bech for his description of the dames de voyage, Danaher models how anxieties about legitimacy have come to contort the story of the sailors' dolls. This anxiety is professional, but it is also deeply cultural—and, in particular, highly gendered. Citing Ferguson instead of Bech means citing a man instead of a woman and thus, by extension, suggesting that citing a man's writing imparts more legitimacy than citing a woman's. The gender politics of this move are also tied to the contents of the pieces of writing in question. I suspect that one reason that Danaher leapfrogs over Bech in his citation to Ferguson is that Bech explicitly frames her discussion of the dames de voyage as part of a critical feminist history—one that casts serious doubts on the ethics of present-day sexual technologies like sex robots. By contrast, Danaher's essay and the larger collection in which it appears argue for the ethical acceptability of sex with robots, demonstrating a vested interest in sidelining objections to sexual technologies.[59] In this context, citing a feminist critique of sex dolls would threaten to delegitimize the project of legitimizing sex with robots in the first place.

Other examples from this most recent wave of publications demonstrate how bids for legitimacy performed through citational practices have taken many forms. The 2016 *Penthouse* article about the history of sex dolls, for example, draws from Ferguson but declines to include any sources at all.[60] We can see this as an attempt to establish a kind of erotic legitimacy. In the context of *Penthouse*, the economy of legitimacy operates differently than in the context of an academic publication. Here, in the pages of a magazine full of pornographic photographs, citations disrupt legitimacy rather than bolstering it. To be legitimate in this context, a piece of writing must offer interesting anecdotes without disturbing the flow of sexual perusal. Facts about sex and sexual histories must remain simply facts—that is, statements made without concern for sources or justifications—lest the fantasies that the magazine offers themselves be brought into question. If sex needs citation, it ceases to be sexy, both because it becomes stodgy and because it reveals itself to be fundamentally constructed, historically rooted, and therefore up for question and open to critique.

Making an Origin Story "Original"

If we trace this citational trail back from this most recent microgeneration of texts, into the ones that preceded it, we find that the evolution of the tale of the dames de voyage has been shaped not only by concerns about legitimacy but also by concerns about originality. As mentioned, Ferguson's *Sex Doll* is, in one way or another, often the source for these recent works. So where does Ferguson draw his information from? Despite the fact that Ferguson discusses the dames de voyage at length, he offers only one source to support his version of the tale: a 2006 article titled "Dames de Voyage," credited to author Amy Wolf and reportedly published in a magazine called *avantoure*.[61] His engagement with Wolf's article is brief. After describing the sailors' dolls as rudimentary cloth creations shared between men, Ferguson states, "The dolls could not have been particularly sanitary after multiple uses," an observation he credits to Wolf.[62] Because Ferguson includes no other explanation for his sources, and because the title of Wolf's article ("Dames de Voyage") seems so on point, we might reasonably expect that Wolf's article is itself the source from which Ferguson draws his general facts about the dames de voyage. The reality, as we might by now have come to expect, is far more convoluted.

As with many of the more obscure points in this citational lineage, I found that tracking down Wolf's "Dames de Voyage" in order to compare it to Ferguson's book proved to be a tortuous challenge. The magazine that Ferguson mentions, *avantoure*, was an early internet publication released only as a PDF.[63] Because the magazine and its website are no longer operational, and because the issues were not captured by internet archiving tools like the Wayback Machine, Wolf's article, in its public form, has all but disappeared. This speaks to ongoing issues of technological obsolescence—which have a particularly detrimental impact on cultural artifacts that are deemed too fringe, too morally questionable, or even too illegal (in the case of many sexual histories) to preserve and which often emerge from the very same sorts of artistic and/or sexual subcultures that thrive on the internet. Thus, ironically, the close relationship between sexual technologies and technology as such makes the histories of sex tech particularly difficult to track.

Luckily, I was eventually able to acquire a version of the article through Wolf herself, who provided an earlier draft of the piece (later tweaked by her editor at *avantoure* before publication) during our individual email correspondence.[64] The piece, which Wolf originally titled more broadly "On Water," roves across related topics, offering a series of reflections on the interplays between gender, sexuality, and the sea.[65] Some elements of the article are related to factual histories, while other elements are related to myth and fantasy. Alongside musings on the sirens from Homer's *Odyssey* and the erotic allure of mermaids, Wolf writes:

> Its [sic] recorded that mariners in the 19th and 20th century fleets satisfied themselves with the "Dame de Voyage," a doll infamous for her sloppy seconds, thirds, fourths and hundredths. This makeshift female was historically made of cotton and presumably held together with dried cum. Unsurprisingly, those who used the early, and grossly unsanitary doll were exposed to syphilis and sexually transmitted diseases. In my research I could not find any images of such a doll, which may be for the best. I can only hope the cotton corpses have been tossed overboard and are now contributing the skeleton of a coral reef. Yet, from its popularity on board, the concept of the Dame de Voyage became militarily sanctioned, with the first modern permutation of this sex toy manufactured in Japan and Germany, countries that both boast formidable naval fleets.

Wolf's account, which situates the dames de voyage in the nineteenth and twentieth centuries, seems to offer many notable details. It appears to speak not only to how the dolls were made (out of cloth and "presumably held together with dried cum") but also how they affected the lives of the sailors who used them, such as by spreading sexually transmitted diseases. Wolf also ties these dolls to "modern permutations" supposedly manufactured in Japan and Germany, in some way tied to their own "formidable naval fleets." What stands out most from Wolf's account of the dames de voyage, however, is her speculative tone. Her description of the dolls as "cotton corpses" that have become part of a coral reef is grim but also poetic—recalling the visions of subaquatic societies from Afrofuturist art (see chapter 6).

As a source of concrete historical information, however, Wolf's article proves to be a dead end. It includes no citations or sources. In my correspondence with Wolf, she was unable to recall where she learned about the dames de voyage (understandably, given that her piece was published more than a decade earlier), though she does mention conducting research in the article itself. In my own extensive research, which has included tracking down an exhaustive list of nearly all published works that even briefly mention the dames de voyage, I have not found a source predating Wolf's 2006 article that gives the same details she does. Where she was drawing from when she wrote "Dames de Voyage" remains a mystery. It's possible that some version of the tale may have circulated on the internet in the late 1990s or early 2000s and that she was pulling from an online blog post or message board comment that has itself since been lost—yet another example of the ways that internet technologies are bound up with the history of sex and how that history is told.

Yet the trail of clues does not go cold with Wolf. Rather, if we return to Ferguson's *Sex Doll*, we find that he too was drawing from an uncredited text. Wolf was not, in fact, his main source of information for his description of the dames de voyage. The facts that appear in Wolf's and Ferguson's texts notably differ. If we look closely, we find that Ferguson's account actually closely mirrors the one provided in another popular history published three years earlier: David Levy's *Love and Sex with Robots* from 2007. At first glance, *Love and Sex with Robots* seems an unlikely candidate for the shadowy text behind Ferguson's work. Here, once again, the facts do not match. Ferguson and Levy place the dolls roughly two hundred years apart and attribute them to sailors from different European countries. However, both Ferguson and Levy run through a strikingly similar accounting of the early cultural life of the sex doll—moving from the dames de voyage to a discussion of sex dolls in sexologist Iwan Bloch's 1907 book *Das Sexualleben unserer Zeit* (The sexual life of our time) to an account of a complicated mechanized sex doll supposedly described in René Schwaeblé's 1904 text *Les Détraquées de Paris* (The wild women of Paris) and then to Henry Cary's 1922 book *Erotic Contrivances*.[66] Clearly, Levy's navigation through these sources and the narrative of history they present has offered a roadmap for Ferguson. Still, curiously, Ferguson does not cite Levy in his writing on the dames de voyage, though he does cite *Love and Sex with Robots* elsewhere in his book.[67]

This obfuscation can be understood as an attempt to make the research presented in Ferguson's *Sex Doll* seem more original and less similar to Levy's. Yet, in an attempt to downplay these similarities, Ferguson ends up not only skipping a citation to *Love and Sex with Robots* but also scrambling much of the information about the dames de voyage found in Levy's book. All of the elements of Ferguson's account of sailors' dolls do appear somewhere in Levy's text, but in creating his own account, Ferguson has brought together snippets from various parts of *Love and Sex with Robots*. From Levy's discussion of the dames de voyage themselves, Ferguson pulls the idea that the dolls were made of cloth and used by sailors to quell sexual frustration on long voyages. From a later point in Levy's book, in which Levy talks about Dutch wives (which Levy describes as early leather dolls made by the Japanese and adopted by the Dutch), Ferguson pulls the idea that the dolls were used beginning in the seventeenth century by Dutch sailors. In this way, Ferguson inadvertently creates a new version of the tale, one that is then passed down through the subsequent texts that draw from his book.

Ferguson is not alone in altering the tale through efforts to establish originality. The greatest point of such distortion comes from Levy's *Love and Sex with Robots* itself. On the surface, Levy appears to have performed a significant amount of primary research to support his description of the history of sex dolls. His writing on the subject features lengthy quotes from a number of seemingly historic sources, including writing by Bloch, Schwaeblé, and Cary, as well as an 1899 novella titled *La Femme endormie* (The sleeping woman).[68] But these sources also do not add up. Levy describes the dames de voyage as sailors' dolls, yet the texts he draws from—as we will see at greater length in chapter 2—in fact describe elaborate commercial sex dolls reportedly sold in France at the turn of the twentieth century. This point of confusion can be explained by yet another attempt to muddy the citational trail and make the story of sex tech's origins seem more original.

At the very bottom of the rabbit hole, we find the earliest (and itself the most original) work in the network of twenty-first-century texts to tell the tale of the dames de voyage: Cynde Moya's 2006 dissertation, "Artificial Vaginas and Sex Dolls: An Erotological Investigation."[69] Levy's own dissertation, originally titled "Intimate Relationships with Artificial Partners," was completed in 2007, a year after Moya's; Levy revised his dissertation to become the book *Love and Sex with Robots*.[70] Moya's dissertation, by contrast, was never published as a book. However, Moya did share her dissertation with Levy while he was in the process of writing his own.[71] While Moya's dissertation has a different focus than Levy's, offering a history of sexological writing about fornicatory dolls rather than a treatise on sex robots, the similarities between certain sections of the two projects are striking. The whole motley lineup of primary sources that Levy uses to support his claims about the dames de voyage appears presented in a nearly identical way as in Moya's project, without citation to Moya. What is more, some particularly nitpicky sleuthing suggests that Levy does not appear to have read many of these texts in full but rather simply drew passages from Moya's dissertation.[72]

The case of Levy's and Moya's respective works serves as a particularly clear example of how the bids for originality that have shaped the contemporary evolution of the tale of the dames de voyage are themselves highly gendered. Levy (whose expertise lies in artificial intelligence) declined to cite the research of Moya (whose expertise lies in the history of sex) precisely on the matter of sexual histories. Across this lineage, a worrying pattern emerges of men drawing from the historical research of women without giving them proper credit. Thus, the gender politics of the tale of the dames de voyage lie not only in the contents of the story itself, nor in how the story is being used, but also in how the story itself has formed. This imaginary also seeks to rewrite the origin story of such scholarship itself, positioning men as the researchers who have drawn together original insights and writing women not only out of the history of sexual technologies, but also out of the work of uncovering and telling that history.

It is in this act of blurring his sources that Levy creates the very tale of the dames de voyage as we know it today. For her part, though she does mention the dames de voyage, Moya never describes them as sailors' dolls, nor does she locate the dames de voyage anywhere within history before the 1900s. Yet Levy, like Ferguson after him, jumbles different elements from Moya's work—such as by transposing a reference in Moya's dissertation to "seamen's brides," which she describes as blow-up dolls used in sex shows during World War II, onto the dames de voyage, and morphing the reference to become a new name for the dames de voyage: "sailors' sweethearts."[73] What we find, when we follow this citational trail across the first two decades of the twenty-first century, is that most contemporary accounts of the dames de voyage can be traced, in one way or another, back to Levy. And Levy's account can be traced to a distorted repacking of research by Moya.[74] Thus has the tale of the dames de voyage, in its twenty-first-century iteration, been born.

These effects are nowhere more apparent than in the image often presented as a picture of the dames de voyage, which is included in *Love and Sex with Robots* (see figure 0.2). Although Levy does not label the dolls as dames de voyage per se, his placement of the image strongly suggests that it represents the sailors' dolls.[75] Indeed, if we look back to Levy's dissertation, we find that before Levy revised his project to publish it as a book, he labeled the image as "dolls in the female form, most often made of cloth and used as sexual outlets by sailors on board ship."[76] In both his book and his dissertation, Levy does credit Moya as his source for this image.[77] It appears in Moya's dissertation as well—but with some key differences.[78] When Moya presents the image, she admits that she has not been able to translate the German text written on it and therefore does not know what it represents.[79] She speculates that it might be "a wonderful drawing of [the type of] sex doll . . . to which Bloch could be referring" in his discussion of early 1900s elaborate commercial sex dolls.[80] In addition, Moya explains where she found the image, pointing to a set of visuals published as a supplement to the 1927 book *Ergänzungswerk zur Sittengeschichte des Lasters* (Supplement to the moral history of vice).[81] Thus, Moya provides both caveats and context for the image, pondering whether it might be a dame de voyage while leaving a citational trail for others to follow.

By contrast, when Levy picks up the image, these nuances are lost. Levy interprets the dolls in the image as dames de voyage—and so, for the purposes of subsequent authors who pull from Levy, they have *become* the dames de voyage. In reality, however, the image that both Moya and Levy replicate is not a set of photographs of sailors' dolls at all. Nor is it a set of photographs of mechanized dolls made from rubber. Nor even, as Moya also suggests, is it an advertisement for a doll available for purchase. Rather, as the German captions on the original image make clear, it is a set of illustrations of a sex doll made by a prisoner—a fact that I return to when discussing sex dolls made in prison in chapter 7. For now, it suffices to note how this image, in trading hands through twenty-first-century texts, has drastically changed meaning: from an artist's rendering of a prisoner's doll to an accepted visual representation documenting the supposed existence of the fabled sailors' sex dolls. Continuing the metaphorical game of telephone, the tale of the dames de voyage comes to life as "facts" about the dolls and even images of them are shuffled down the line, whispered and garbled: sometimes by accident and sometimes, seemingly, very much on purpose.

Questioning Histories of Sexuality and Technology

Mapping the network of contemporary texts that tell the tale of the dames de voyage reveals that the story of the "very first sex doll," in its present-day manifestation, is far more complex and multifaceted than a simple historical anecdote. With its many contradictory elements and underlying cultural anxieties, these tellings of the tale raise more questions about the actual history of sexual technologies than they provide answers. As for the work of tracing the citational lineage behind the contemporary tale, the results offer insights of a different sort. What we find in mapping out the often-implicit connections between authors is that the story of the sailors' dolls is not one that has simply passed from source to source. Instead, it has mutated and evolved, taking shape through retellings. These twists and turns have done more than alter this tale: they are precisely what have made the tale. The dames de voyage, as we know them today, do not exist separately from this shifting, contradictory web of stories about them.

We are left then to wonder: If none of these contemporary accounts are fully factual, what is the truth about the dames de voyage? Over the next two chapters, I continue to search for the sailors' dolls, delving into texts and archives that predate present-day discussions of sex tech and its origins. However, before moving on, I want to linger for a moment on the broader implications of how the contemporary tale of the dames de voyage has formed. The story of the sailors' dolls stands as evidence of how historical anecdotes are not told but *made* and how the act of writing and publishing on subjects of history also entails the act of making history. These may seem banal observations for those many scholars who are already sold on the notion that history is both an imaginary construct and an imperfect art. Yet when it comes to both the history of sex and the history of technology, this willingness to understand history—all history—as in some sense imagined is still underdeveloped. Nowhere is history more vulnerable to misrepresentation, nor richer with potential for reimagining, than when it is the history of how people have sex with, through, and alongside tech. In this sense, the contemporary tale of the dames de voyage serves as a provocation: to question what we think we know about the histories of both sexuality and technology and to see these histories as an invitation to turn a critical eye toward the present as well as the past.

2 How Fantasy Became History: The *Dames de Voyage* in Pseudoscience, Erotica, and Advertising

Contemporary authors, it turns out, are far from the only ones to daydream about the *dames de voyage*. The twenty-first-century story of the dames de voyage as sailors' dolls may be a fabricated amalgamation of historical elements, as I address in chapter 1, but references to the dames de voyage (whatever they might be) go back much earlier. Long before treatises on the origins of sex tech or present-day attempts to justify the supposed rise of sex robots, the dames de voyage were making appearances in a variety of works. Sometimes they were the subject of scholarship; sometimes they were being advertised for sale; and sometimes they took center stage in flights of the pornographic imagination. Today, the dames de voyage are described as having been constructed by sailors and are positioned in the role of "the very first sex doll." Yet, previously, in the twentieth century and the later decades of the nineteenth, sex dolls referred to as *dames de voyage* took on very different forms, often in much more explicit erotic fantasies about possibilities of sex with tech.

If we begin to work our way backward from the present in search of the dames de voyage, these dolls quickly reveal themselves to be something very different than expected. I kicked off the search for the dames de voyage in the previous chapter by looking at contemporary sources: texts published over the last two decades that tell and retell the story of the sailors' dolls. My search led me down a winding citational trail, eventually back to writing by David Levy and research by Cynde Moya, both of whom drew their information from a shared set of early twentieth-century sources that seemed to confirm the existence of the dolls, possibly as makeshift sailors' dolls and possibly as commercially produced dolls with complex interior workings. By tracing this lineage, I found that in its present-day forms, the tale of the dames de voyage has largely been the product of contortions and obfuscations as various accounts of the origins of the sex doll have been passed from author to author. That does not mean, however, that the dames de voyage did not exist in one form or another. Were they really rudimentary sex dolls made by European sailors? Or were they elaborate mechanized human replicas—predecessors to the sex robots of today and tomorrow? Or were they something else entirely?

In this chapter, I move past the present moment, stepping backward through a longer and older history of references to sex dolls and specifically the dames de voyage. The works considered here span a range of time periods and locales, from the late 1960s and early 1970s in the United States to the 1900s, 1910s, and 1920s in Germany and to the 1890s in France. They represent an eclectic array of materials, including pseudoscientific pulp paperbacks designed to titillate, sexological treatises that claim to comprehensively catalog unusual human sexual practices, erotic fiction about sex between humans and dolls, and boastful advertisements for the latest in fin de siècle sexual technologies. However, the process of searching for the dames de voyage in these earlier works also serves as a window into something much bigger than evidence of the dolls' existence. It lays bare the role of fantasy in the making of historical fact. In exploring and critiquing these texts, this chapter demonstrates how works of fiction—whether in the form of fake sociological studies, pornographic stories that parody popular novels, or the flowery promises of sex toy advertising—have been taken up as primary documentation in the construction of what is, at best, a questionable history. At times, these works have been misinterpreted. At other times, they have been taken at face value, allowing euphemism and thin veneers of science to pass as researched fact. Perhaps most interesting are the times in which we find that the fictional nature of a text was clear, and yet those citing it chose to read it as true: a reflection of a willful desire to see the stuff of libidinous longing become the substance of history.

All of the works discussed in the following sections, from erotica about sex doll orgies to catalogs for sexual novelties, sell a fantasy of one kind or another. It is these fantasies, more so than any particular material or cultural reality, that have formed the lineage that brings the tale of the dames de voyage up from the end of the 1800s into the present day. Yet the materials presented here are not simply false, ripe for debunking. Instead, precisely through the fantasies they offer, they show us elements of history that are often underconsidered and yet very much real. They demonstrate, for example, how the imaginaries that surround sex tech in the twenty-first century have repeated over time. It turns out that the dames de voyage have long been an important node in conversations about the sexual possibilities of technology. Simultaneously, this genealogy shows us how histories of sex, histories of technology, and history more generally are made through (sexual) excitement and (sexual) desire. Which works we choose to read as true, it seems, depends on which visions of history we long to write into reality.

Pseudoscientific Pulp: The *Dames de Voyage* in the Mid-Twentieth-Century Sex Tech Craze

Before the works of the 2000s, there is a considerable jump back in time to the previous generation of texts that reference the dames de voyage.[1] Moving backward from the present day, the first period we find ourselves in is the 1960s and 1970s, amid a wave of pseudoscientific texts about sex and sexual histories written by American authors.[2] These texts, mostly books aimed at a mainstream readership, typically present themselves either as scholarly works offering sociological insights into various sexual practices (playing off the notoriety and academic aura of the famous Kinsey reports) or as how-to manuals for readers interested in the use of various sexual contraptions.[3] Such works are, however, best understood as thinly veiled examples of erotic pseudoscience: books that aim to tantalize rather than educate, capitalizing on the cultural moment of America's "sexual revolution" by describing a dazzling array of sex acts and sexual devices with little actual engagement with research or history. It is also notable that the consumer landscape for sex dolls themselves was changing at this time, with a growing number of advertisements in English-language publications for products like the "inflatable play-girl" (see figure 1.1) and the "loveable doll" ("Wonderfully human details! Ready for exciting action!").[4] Thus, as the place of sex dolls was changing in the American market—bringing them, by extension, more prominently into the American popular consciousness—we also see a spike in texts that claim to tell the cultural history of these dolls.[5]

Such books themselves stand within a long lineage. They can be seen as playing off of older traditions of protopsychoanalytic and sexological

writing, such as Richard von Kraft-Ebbing's *Psychopathia Sexualis* from 1886
or Havelock Ellis's *Studies in the Psychology of Sex* (circa roughly 1897–1930).
These earlier works helped establish the practice, still common, of writing
about sex as the work of list-making: enumerating and cataloging lists of
sexual proclivities and "pathologies," a vision of sexual epistemology as
encyclopedic in nature rather than deeply engaged with culture or history.[6]
Midcentury texts of this sort take up the notion of sexual scholarship and
reimagine it as sexy, repackaging erotic thrill in unassuming volumes that
pass as legitimate on a home bookshelf. (It is likely not coincidental that many
of my copies of these books, which I have obtained through unrelated sellers, smell strongly of cigars. I like to imagine that, in their former lives, they
sat on the shelves of various dark, plush personal libraries where their owners smoked and browsed "scientific" volumes.) Many of these texts draw
from each other, with or without citing one another. The proliferation of
such similar works—which seem to have been released so rapidly that, in at
least one case, books on the same topic with identical titles were written by
different authors and published in the same year in the same city—speaks
to the popularity of this then-emerging genre that Cynde Moya describes
as "pseudo-scientific pulp."[7]

Consider, for example, the 1970 book *Sex Devices and How to Use Them*,
written by Amy and Ashton Dumont, which contains a chapter on fornicatory dolls.[8] The book opens by positioning itself as a legitimate scholarly
work, stating that "the intention of the authors and the researchers of this
study is to reach . . . the individual who wishes to follow the sociological
and scientific patterns involved with our topic [or] the interested student of
sexology."[9] As for the actual contents of the book, however, they tell a different story. *Sex Devices and How to Use Them* advertises itself on its cover as
"photo illustrated," and indeed nearly every other page is accompanied by
a pornographic photograph, mostly of group sex or objects being inserted
into orifices—images which, it should be noted, rarely bear any illustrative
relation to the subjects of the chapters they accompany. Even the text of the
book itself constitutes little more than pornographic anecdotes presented
as firsthand accounts and lightly packaged in the rhetorical framing of
research. The chapter on fornicatory dolls, for instance, consists mainly
of an American soldier's story about watching a sex show in Berlin during
World War II that involved a blow-up doll, as well as another story—a sort
of Bluebeard-meets-sex-dolls fantasy—about a man who housesits for a

wealthy friend only to discover a locked room full of "plastic playmates," who become participants in ensuing orgies.[10]

It is in texts such as this that we find this first precontemporary cluster of references to the dames de voyage. For example, writing in his 1969 book *The Humor and Technology of Sex*, Paul Tabori states, "We have seen in our earlier sections [of the book] how the masturbatory demands for artificial penis and vagina created an art and an industry. In the twentieth century the latter have evolved into the so-called *dames de voyage*. . . . Not only the genital region but the whole feminine body is reproduced from various materials, and even its secretions are imitated by small quantities of oil."[11] Tabori's text is one of those earlier works often referenced by a number of more recent, twenty-first-century authors who tell the tale of the dames de voyage as the story of sailors' sex dolls. For this reason, it is interesting to note that Tabori describes the dames de voyage not as makeshift dolls constructed at sea but rather as full-body sex dolls equipped with additional features, such as the ability to release oil to replicate vaginal fluid.

Other accounts from this mid-twentieth-century period present a similar picture. Arguably the best-researched of these texts is *The Illustrated Manual of Sexual Aids*, published in 1973 and written by Evelyn Rainbird.[12] Rainbird states that "vagina substitutes were certainly manufactured in the nineteenth century, if not earlier. . . . In France, 'love dolls' were sometimes accorded the discreet name *Dame de Voyage*, or travelling companion."[13] Later, Rainbird mentions a "contemporary" (i.e., dated 1972) catalog of sex toys that includes a "crudely molded torso with an even cruder painted representation of pubic hair around the vaginal aperture," which she refers to as a "rubber *Dame de Voyage*."[14] For Rainbird too, then, it seems that the term *dames de voyage* broadly signifies penetrable sex dolls made to look like women, whether as full bodies or mere torsos, rather than the sailors' dolls typically described in more recent texts.

What is telling about these pseudoscientific works from the 1960s and 1970s, in addition to how they describe the dames de voyage, is the way they situate these descriptions. Technology emerges as a central concern in this wave of "scientific" pulp. We can see this evidenced in many of the titles mentioned so far, as well as others, which trade in terms like *sex devices*, *sex gadgets*, and *technology of sex*.[15] In such works, a fascination with sex merges with a fascination with technology. The sexual technologies addressed in these books also extend beyond purpose-made sex toys and into the realm of other mid-twentieth-century items supposedly repurposed for sex. This includes even domestic technologies like the portable vacuum cleaner, as we see provocatively visualized on the cover of the 1972 book *A Housewife's Guide to Auto-Erotic Devices in the Home,* where a woman throws her head back in ecstasy as she presses a suction hose to her nipple (figure 2.1): a literalization of the sense that this period of "sexual revolution" was itself bound up with a shifting relationship between consumers and technology.[16]

The fascination with technology that recurs in these texts provides us with new insights on two fronts. First, it shows that long before the tale of the dames de voyage became wrapped up in twenty-first-century histories of high-tech sex, references to sex dolls were already being couched within a broader interest in the interplays between sex and technology. That is, for at least the last fifty years (and, as we will see in a moment, even longer), sex dolls themselves have been understood both in conjunction with and indeed *as* technology. Second, this emphasis on the erotic allure of technology, gadgets, and devices that emerges within these texts shows us that the fascination with sex tech that is prevalent today itself has its own cyclical history. This is not the first time that America has been hit with a wave of fascination and hype—and accompanying moral panic—about how changes in technology and culture are suddenly making possible exciting new ways to have sex through tech.

The *Dames de Voyage* in German Sexology: Sex Robots Made from Pneumatic Tubes

Before pseudoscience, in the historical trajectory of the tales of the dames de voyage, came "science" itself. The precursors to these mid-twentieth-century pseudoscientific texts, from which the pulp books of the 1960s and 1970s themselves took many of their cues, came in the form of yet another wave of published works: early twentieth-century sexological studies. Although they rarely provide sources (Tabori's and Rainbird's books are exceptions), many of these mid-twentieth-century American texts draw information from this earlier generation of writing about sex that is more firmly intellectual, though not without its own dalliances in titillation. There are some examples of these sexological works that emerge from the American context, such as Henry Cary's 1922 book *Erotic Contrivances:*

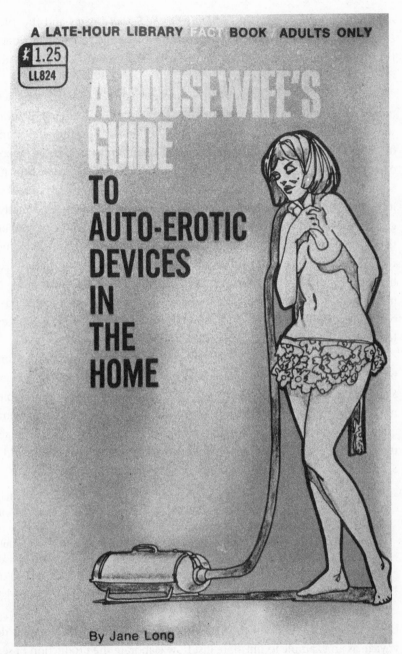

A LATE-HOUR LIBRARY FACT BOOK   ADULTS ONLY

1.25
LL824

A HOUSEWIFE'S GUIDE
TO
AUTO-EROTIC
DEVICES
IN
THE
HOME

By Jane Long

Figure 2.1
Cover of the 1972 book *A Housewife's Guide to Auto-Erotic Devices in the Home*, exemplifying the pulp pseudoscience of the 1960s and 1970s. *Source:* Greenleaf Classics.

Appliances Attached to, or Used in Place of, the Sexual Organs and the 1961 *Encyclopedia of Sexual Behavior*, edited by Albert Ellis and Albert Abarbanel.[17] However, to find most of these works, we need to look elsewhere—to the traditions of German sexology.

Sexology is the scientific study of human sexuality. Americans are likely to be more familiar with later figures like Alfred Kinsey or Virginia E. Johnson and William H. Masters, but in fact the field of sexology has many of its roots in Germany.[18] The intellectual community of German sexologists took shape in Berlin in the 1910s and 1920s, with the Institut für Sexualwissenschaft (Institute for Sexology) run by Magnus Hirschfeld at its center. Widely considered a hub for the explorations of non-normative gender and sexual identities that briefly flourished in the Weimar Republic, the work of the Institute for Sexology came to a violent end in 1933 when the institute and its archival holdings were destroyed by the Nazis.[19] For historians of sex, the destruction of Hirschfeld's institute represents a site of unknowable loss. The materials contained in its archives, which could have offered invaluable documentation of pre-twentieth-century sexual cultures, has become the stuff of speculation and mourning.[20]

Nonetheless, many sexological treatises from the period survive—a number of which mention the dames de voyage. Leo Schidrowitz's 1927 book *Sittengeschichte des Lasters die Kulturepochen und ihre Leidenschaften* (Supplement to the moral history of vice)—from which Moya drew the image now often misinterpreted as representing sailors' sex dolls, as described in chapter 1—is one such work.[21] In a section titled "Fetishism and Its World," Schidrowitz writes:

> The use of phallic implements by women and that of vaginal replicas by men, insofar as they are used to masturbate, lies less in the fetishistic field than in that of masturbation. Every imaginable material has been used, at one time or another, to produce such phalluses, in order to adapt them as closely as possible to nature. This "branch of industry" has accommodated male sexual fantasies all the way up to the *"dames de voyage."* These are replicas of entire female bodies, not just the lap area, with lifelike genitals, the secretion of which is even imitated by oil.[22]

Here, writing in Europe in the 1920s, Schidrowitz presents the dames de voyage as an example to illustrate just how elaborate the production of sex toys had supposedly already become. Describing the dolls as replicas of the "entire female body," he presents them as the crowning achievement of attempts by the "industry" (it is intriguing to think not just about sex toys of yore but also sex toy *industries* of yore, a topic explored in chapter 4) to "accommodate male sexual fantasies."

Schidrowitz's text is one of the latest in this particular German sexological lineage. Before Schidrowitz, the trail of the dames de voyage gets briefly murky. In *Sittengeschichte des Lasters*, Schidrowitz cites a 1910 text by Erich Wulffen, *Der Sexualverbrecher* (The sexual criminal), as his source for information about the dames de voyage.[23] Indeed, Wulffen does discuss the dames de voyage in similar terms, presenting the dolls as one point in a larger consideration of the "ethnological phenomenon" of sex with objects.[24] Another book from the same year, Georg Back's *Sexuelle Verirrungen des Menschen und der Natur* (Sexual aberrations of man and nature) similarly describes the dames de voyage, positioning a discussion of sex dolls after a retelling of the myth of Pygmalion, a common rhetorical pairing across texts that discuss sex dolls.[25] Yet if we look beneath the surface, we find that Wulffen, Back, and other authors writing sexological texts in this period are all in fact drawing information about the dames de voyage from the same source: the writing of Iwan Bloch.[26]

Sometimes referred to as the "father of sexology," Bloch was a historian from Berlin who proposed the creation of a new "sexual science" that would include "cultural, social, and historical studies alongside biological and psychological findings."[27] Arguably his most influential book, and the one that circulates most widely outside of the German-language context, is his 1907 *Das Sexalleben unserer Zeit* (The sexual life of our times).[28] Here, Bloch includes a passage that has become foundational for the circulation of accounts of the dames de voyage over the last 110 years, shaping both early sexological references to the dolls and continuing to appear in twenty-first-century texts.[29] Following a description of Parisian brothels where "pygmalionists" could reportedly pay to view sex workers posing as nude statues of Roman goddesses, Bloch states the following (I quote this passage at length because both its "factual" contents and its citations are important):

> In this connexion *[sic]* we may refer to fornicatory acts effected with artificial imitations of the human body, or of individual parts of that body. There exist true Vaucansons in this province of pornographic technology, clever mechanics who, from rubber and other plastic materials, prepare entire male or female bodies, which, as *hommes* or *dames de voyage*, subserve fornicatory purposes. More especially are the genital organs represented in a manner true to nature. Even

the secretion of Bartholin's glands is imitated, by means of a "pneumatic tube" filled with oil. Similarly, by means of fluid and suitable apparatus, the ejaculation of the semen is imitated. Such artificial human beings are actually offered for sale in the catalogue of certain manufacturers of "Parisian rubber articles." A more precise account of these "fornicatory dolls" is given by Schwaeblé ("Les Détraquées de Paris," pp. 247–253). The most astonishing thing in this department is an erotic romance ("La Femme Endormie," by Madame B.; Paris, 1899), the love heroine of which is such an artificial doll, which, as the author in the introduction tells us, can be employed for all possible sexual artificialities, without, like a living woman, resisting them in any way. The book is an incredibly intricate and detailed exposition of this idea.[30]

Here, Bloch situates the dames de voyage within a discussion of "artificial imitations of the human body." Notably, he mentions both men's and women's bodies: *hommes* or *dames de voyage*. Specifically, his interest lies in particularly elaborate versions of such items, which he says are being constructed from a variety of materials, complete with pneumatic tubes to secrete faux bodily fluids, by "true Vaucansons" (a reference to Jacques de Vaucanson, the famous French inventor of early automata), who sell these works of "pornographic technology" through catalogs advertising rubber items. Bloch also points toward additional sources that seem to document the existence and workings of these "astonishing dolls."[31]

There are a number of key things to note in this passage from Bloch— which has been, in subsequent years, both much quoted and much cribbed from in the writing of others. The first is Bloch's own fascination with the dolls as technological. He calls the people who build them "clever mechanics" and compares them, implicitly, to inventions that now hold a celebrated place in histories of science and technology, like Vaucanson's 1739 mechanical Canard Digérateur (Digesting Duck).[32] It is perhaps unsurprising then that twenty-first-century authors have positioned the tale of the dames de voyage (even in its version as the story of sailors' rudimentary sex dolls) within narratives of technological innovation, given that Bloch himself seems enthralled with the idea of the dolls' "possible sexual artificialities." Relatedly, it is also worth observing the dreamy, aroused quality of Bloch's description, which starts out matter-of-factly and ends in a loose reflection on an "incredibly intricate and detailed exposition" of what it might be like to have sex with dolls who cannot reject one's advances. Indeed, this is a trope that recurs across various historical moments and iterations of the tale of the dames de voyage, in which the dolls' imagined inability to resist is often figured as one of its allures.[33] Thus, the problematic consent politics that continue to characterize the cultural life of sex dolls and sex robots can already be seen forming in much earlier works.

However, there are also elements of this passage from Bloch that notably do not match up with later accounts. For instance, Bloch clearly mentions both dolls shaped like (cisgender) men and those shaped like (cisgender) women. He also describes how apparatuses contained within the dolls supposedly secreted fluids replicating either vaginal mucus or semen. Yet most later accounts of the dolls, including those that directly draw from Bloch, sidestep the notion of male sex dolls, instead focusing only on the *dames* de voyage rather than their *hommes* counterparts and repeating only Bloch's description of the tubes filled with oil that replicate "the secretion of Bartholin's glands," glands located near the entrance to the vagina, rather than those that purportedly produce a semen-like substance. As I will mention again in chapter 5, it seems that part of transforming the history of sex tech into a story about straight men's desires has entailed writing sex dolls shaped like men out of that history—whether because the notion of hommes de voyage worrisomely renders men themselves into objects or because the image of the male sex doll comes with a homoerotic valence, stirring up homophobic anxieties that the history of the sex doll might, in fact, be in part a queer one.

The most crucial distinction to note in Bloch's description of the dames de voyage is actually also the subtlest one. Almost all of the twentieth-century texts described in chapter 1 that draw from Bloch (whether Bloch's claims contradict their depiction of the dames de voyage as sailors' dolls or not) interpret this particular passage to mean that the elaborate mechanical dolls sold through Parisian rubber goods catalogs that Bloch is describing were *one and the same* as the dames de voyage. That is, they take Bloch's text to mean that the dames de voyage were themselves, at the start of the 1900s, protorobotic full-body contraptions. Yet, if we look more carefully, we see that Bloch is in fact saying something else. He writes that the dolls in question are being used "as *hommes* or *dames de voyage*." The as here is key.[34] These dolls are not themselves quite the same thing as dames de voyage, at least not as they were known in Bloch's time. Rather, dames de voyage seems to signify something broader, more equivalent to *sex doll* or a sex toy for insertion that resembles a part of the human body.[35] Thus, a more careful reading of Bloch suggests that these elaborate "artificial imitations of the human body" were not literally the dames de voyage; they were simply one type of item within a broader category.

In this line of sexological works, we see once again how shifting notions of the dames de voyage have been shaped, across history, by slippages and confusions between sources. Yet a new picture also emerges through these works. Perhaps, if these earlier texts are to be believed, the dames de voyage were not the sailors' dolls described in the twenty-first century. But maybe the complex technological creations described by Bloch *did* exist: true ancestors to the sex tech of today. To explore this possibility, we must look to Bloch's own sources—those texts he points to as proof of the "astonishing" sex dolls purportedly available for purchase in the first decade of the twentieth century, moving backward yet again to a previous generation of works.

From Erotic Fiction to Historic Fact

Luckily, Bloch is clear about the sources he uses to support his description of the dames de voyage. The first is a text by an author whose name Bloch gives as Schwaeblé, titled *Les Détraquées de Paris* (translated roughly as The wild women of Paris or The crazy women of Paris). Bloch says that this text offers a "precise account" of the complex fornicatory dolls supposedly available for sale.[36] The second text, this one by someone called Madame B., is titled *La Femme endormie* (The sleeping woman); Bloch describes it as an "erotic romance" that offers an "incredibly intricate and detailed exposition" of the idea of how a doll might be used "for all possible sexual artificialities."[37] Lastly, Bloch references "catalogs of certain manufacturers of 'Parisian rubber articles,'" stating that these catalogs include listings for "artificial human beings" whose bodies can be made to secrete fluids.[38] Each of these sources appears, at first glance, to corroborate Bloch's claims about the dolls: that they were available for sale, that they had impressively elaborate inner workings, and that they were being put to use by eager customers. These same texts are also often referenced in contemporary writings about the dames de voyage and the history of sexual technologies more broadly, where they form a backdrop of (what seem to be) firsthand accounts confirming the origins of the sex doll.

However, if we locate and dig into these texts themselves, we find that they are not firsthand accounts at all. Far from being sexological works or factual primary sources, all three of these sources are clearly the stuff of fiction and fantasy. Consider Schwaeblé's *Les Détraquées de Paris*. This work, originally published in 1904, is in fact a collection of twenty-seven erotic short stories.[39] Although the book carries the subtitle *Étude documentaire* (Documentary study), the text itself is anything but scholarly. Instead, featuring an array of tales about different sexual encounters and proclivities, it entices its readers with the promise of a delightfully voyeuristic glimpse into the titillating underbelly of Parisian life—and predominantly into the lives of young, beautiful, debauched Parisian women who have sex with other women.[40] The stories in Schwaeblé's text come with coy titles like "Smokers of Opium," "Milk Baths," and "Women and Beasts" (which features a monkey—and, yes, the story goes exactly as you might imagine). The volume boasts an array of black-and-white *illustrations d'après nature* (realistic illustrations), a common feature of pornographic works from the time, which range from the sexually suggestive to the sexually explicit. Flipping through the pages of the volume reveals image after image of women dancing bare-chested, embracing lovers in their boudoirs, gazing hungrily at each other's partially naked bodies (figure 2.2), and even fencing one another in semitransparent, form-fitting costumes (figure 2.3).

If we need further proof that Schwaeblé's book is the stuff of fantasy rather than sociological reporting, consider how *Les Détraquées de Paris* was advertised around the time of its release. In 1907, an ad for the book appeared in a publication called *Paris-Fêtard: Guide secret de tous les plaisirs* (Partying in Paris: A secret guide to every pleasure).[41] A handbook for sex tourism, *Paris-Fêtard* boasts information about local prostitutes, brothels, "artistic cabarets," and more. Here, Schwaeblé's book is listed for sale among a collection of other "curious volumes," alongside similarly tantalizing, pseudoscholarly titles like *The Modern Orgy* and *The Studio of Debauchery*.[42] In 1910, another advertisement for the book appeared in the back pages of the humor newspaper *Le Rire: Journal humoristique* (Laughter: Comedy newspaper), where *Les Détraquées de Paris* was promoted as one among many "lovely illustrated volumes" for sale.[43] Other books on offer included *The Love Societies of the 18th Century*, *Adulterous Women*, and—my personal favorite—*Their Panties (Feminine Indiscretions)*. Surely this is not the context in which we would expect to find a purely nonfictional book.

The most damning evidence that the contents of Schwaeblé's text should not be misconstrued as a factual account lie in the text itself. Of

Figure 2.2
An erotic illustration from René Schwaeblé's 1904 book *Les Détraquées de Paris* (The wild women of Paris), which includes a story about an artisan who makes bespoke sex dolls.

Figure 2.3
Illustrations from *Les Détraquées de Paris*, like this one of women fencing, reflect the book's queer overtones.

the twenty-seven stories in *Les Détraquées de Paris*, sex dolls are the subject of only one: a brief vignette, less salacious than some others, titled "Homunculus."[44] Far from being a "precise account" of mechanized dolls sold through Parisian rubber goods catalogs, as Bloch suggests, the story is about an individual artisan who makes highly detailed, bespoke replicas of individual people, like dreamy local celebrities, and sells them at great cost. In the story, a character named Doctor P., who is "famous among occultists" (always a good sign), runs a clandestine business designing artificial men and women.[45] Basic models, the story's narrator tells us, sell for 3,000 francs and take three months of work to create. More carefully rendered models, for which the artist says that he must observe a living subject on multiple occasions, cost 10,000 to 12,000 francs. Even for all that time and expense, the dolls are far from the mechanized wonders that Bloch describes. Schwaeblé's text makes them seem more like wax dolls, with added hair, nails, skin, and teeth.

Even if we believe that there was such an artisan illicitly crafting dolls for the wealthy and lonely of Paris, the details once again do not line up. For one thing, Dr. P.'s business appears to focus more on replicating men than replicating women. Not to mention that the tone of Schwaeblé's story makes the whole account seem like little more than a ribald joke. "Homunculus" ends with a woman patron taking her doll on a "beautiful honeymoon" in Italy by deflating him and packing him into a suitcase.[46] Elsewhere, we do find accounts that suggest that there may have been such one-off shops for creating bespoke dolls in other European cities around this time, though other anecdotes are just as questionable, and neither Schwaeblé's text nor the others paint a picture of the dolls as commercially produced at a sizeable scale.[47] It seems safe to say that though there may be kernels of truth in *Les Détraquées de Paris*, it would be a mistake to take it as an official source for accurate information about the sex dolls for sale in France at the start of the 1900s.

What about Bloch's second source, *La Femme endormie*, the 1899 "erotic romance" that he says offers an astonishing exposition of the construction and use of sex dolls?[48] While Bloch does not refer to *La Femme endormie* as a work of scholarship, he does portray it as a thoughtful, reflective text and a useful source for the scientific study of sex.[49] But *La Femme endormie* is, in fact, nothing of the sort. Instead, it is a work of pornographic fiction, plain and simple, with none of the trappings of the "documentary study" that wrapped Schwaeblé's text in a sociological veneer.[50] Madame B., the story's listed author, is a pseudonym for pornographer Alphonse Momas, a French writer who published more than one hundred erotic novels between 1891 and 1908, using a range of charming names fitting a Sadeian libertine, including Georges de Lesbos, Erosmane, and simply Fuckwell.[51] I am not the only one to recognize the fictitious nature of *La Femme endormie*. Jon Stratton, writing about the trope of "man-made women" across literature and film, characterizes Momas's story as "pornography . . . clearly aimed at a male audience," and explains that the story should be understood as a "carnivalesque mimicry" of other works of fiction that center women created for the pleasure of their inventors, such as Auguste Villiers de l'Isle-Adam's 1886 novel *L'Ève future*.[52] Indeed, far from being a reflection on the actual sexual technologies of the 1890s, *La Femme endormie* is a parody—not unlike modern-day porn parodies, an enduring if curious genre—of more self-serious (yet still implicitly erotic) books about the technological possibilities of men making women.[53]

As for the story itself, its contents are clearly designed to arouse—casting Bloch's description of the work as "astonishing" in a much more lustful light. The work begins in the style of the Pygmalion myth, with a sexually frustrated male protagonist, Paul, who has all but sworn off the company of women, bemoaning his lack of an outlet for his libidinous appetites. A rich Parisian for whom money is no object, Paul decides to hire an artist to build an elaborate doll that can become his sexual plaything: a "will-less creature who would submit ecstatically to his obsessions and his lewdness."[54] When the doll arrives, Paul regards her with smug excitement, describing her as "an admirable woman with very uplifted, very firm breasts, outstanding, appetizing hips, extremely well-shaped buttocks, [and] divinely curved loins."[55] Yet the thing that excites Paul most is not the doll's outward appearance but rather the mechanical complexities of her inner workings. He marvels at length over how various parts of her body can be opened, filled with liquids, cleaned, and made to move with various presses of concealed buttons. Inside, she is composed of a complex network of "wire netting and various tubes going in all directions," which itself holds an array of receptacles for storing warm liquids and catching semen.[56] The story is as indulgent in its descriptions of the doll as a work of technology as it is in its descriptions of Paul's trysts. *La Femme endormie* comes to a graphic climax in a scene of group sex among Paul, the doll (whom he names Mea), the artist who made her, and Paul's new flesh-and-blood girlfriend. It reflects a narrative structure perfectly fitting the logics of pornography, where everyone inevitably sleeps with everyone, even the sex dolls.

First and foremost, this is a story about sex with machines. Bloch describes Madame B.'s "romance" as being about the possibilities of sex with a woman who cannot refuse consent. Really, however, it is a fascination with machinery and mechanization that animates the erotics of *La Femme endormie*. When Paul first opens the crate containing the doll, he undresses her and lays her out on the sofa. Then we read: "His tongue went directly to the artificial cunt and his hands were delicately directed to her buttocks. Fever overtook him. 'Egad! What's this!' he exclaimed after a few seconds. . . . He slid his fingers between her thighs and, dumbfounded, ascertained the presence of a few little whitish drops."[57] These "little whitish drops," we soon learn, are a concoction of milk, vinegar, and egg whites released through a tube near the doll's vaginal opening—substances that Paul must diligently replenish as part of the ritual of opening her up and cleaning her after sex. This ritual is strikingly similar to disassembling and cleaning an engine, and indeed Paul takes special pleasure in operating Mea's body as one might operate a piece of machinery. Ironically, though Paul begins the story by bemoaning the inconveniences of romancing human women, with their "screeching voices" and "gestures of tiredness," he ends up with a lover whom he desires, in part, because she is quite literally *high maintenance*—to use a contemporary term that signals both the (gendered, emotional) "maintenance" of romantic relationships and the (physical) maintenance of machines.[58]

In both *Les Détraquées de Paris* and *La Femme endormie*, we find works of erotic fiction that engage in acts of imagination—fictions that Bloch then takes up (whether erroneously or indulgently) as factual evidence of elaborate dolls actually offered for sale in France at the start of the 1900s. Through these stories, fantasy comes to fundamentally shape the vision of the dames de voyage that Bloch presents. As a result, through Bloch, these same fantasies have formed part of the basis for the vision of the dames de voyage that has been passed down across generations of writing. Yet "documentary studies" and pornographic novels are not the only works of fantasy to feature in Bloch's much-cited text. Fantasy also characterizes his third and final source: advertisements.

Selling the Fantasy of the *Dames de Voyage*

Seemingly the most concrete piece of evidence that Bloch offers for the existence of the dames de voyage—or, at least, items used as dames de voyage—is his reference to catalogs for "Parisian rubber articles." Bloch says that these catalogs offered full-bodied, elaborately constructed "artificial humans" for sale at the time of his writing in 1907.[59] Such catalogs seem to be promising touchstones in the search for the dames de voyage, since we might reasonably expect advertisements to operate in the realm of material fact, describing actual items available for purchase with relative accuracy. Bloch does not go into much detail about the contents of these advertisements. This leaves the question open: If we track them down ourselves, do we find that they indeed serve as proof of the production and sale of elaborate early sex dolls—mechanical creations of "true Vaucansons," as Bloch describes?[60]

While many of the primary materials related to the history of sex tech are difficult to track down, advertisements for sex dolls during this period are particularly elusive. This is in part because, as examples of the ephemera surrounding sex and sex-related industries, the types of catalogs that Bloch is referencing have rarely been deemed valuable for archiving. In addition, some of the items contained in these catalogs, including sex dolls themselves, were of questionable legality at the time of their sale, making the preservation of such catalogs all the more unlikely.[61] Luckily, there do remain a few key archives and entry points for learning about the contents of these catalogs, which were indeed produced at the end of the 1800s and the start of the 1900s.[62] From these we can learn that catalogs of this sort, as well as loose-leaf flyers, were indeed circulated by companies advertising "rubber goods," as well as *appareils speciaux* (special apparatuses) and other items for *usage intime* (intimate usage)—both terms used as euphemisms for sex toys and other sex-related items by a network of sellers operating in Paris starting as early as 1891 (figure 2.4).[63] Existing examples demonstrate that these catalogs typically contained a range of items, most prominently condoms and other contraceptive devices, but sometimes also including dildos, "French ticklers" (cock rings with an extra bit of rubber to stimulate the clitoris), and strap-on harnesses. They also often featured rubber corsets, part of what Manuel Charpy has described as a "craze for rubber clothing."[64]

Figure 2.4
The cover of a 1905 catalog from Société Excelsior, selling rubber goods and other items for "intimate" uses.

Customers interested in receiving these mail-order catalogs—which were illustrated and frequently included "samples," though it is unclear exactly what these were samples of—could request them either for free or for a small cost. Within Paris, the catalogs themselves were advertised through small ads in newspapers and local circulars.

As for listings for sex dolls in particular in these catalogs, those seem to have been quite rare—suggesting that such dolls were only offered through a limited number of sellers. To date, existing scholarship on the history of sex dolls that has discussed ads of this sort has only been able to do so by pointing to two later sources that replicate the contents of one particular catalog advertisement, which is for a *ventre de femme avec vagin artificial* (woman's belly with artificial vagina). These sources are Henry Cary's 1922 *Erotic Contrivances* and Evelyn Rainbird's 1973 *The Illustrated Manual of Sexual Aids*. In his discussion of the dames de voyage, Cary writes, "There is manufactured and sold in Europe today an imitation of the female private parts, even to the pubic hair. . . . Circulars describing them usually call them lady travelers, and recommend them for the use of naval officers and others who are deprived of female society for long periods of time."[65] Relatedly, Rainbird describes how these advertisements, originating in Paris, were also translated and sent abroad. "Such catalogues," Rainbird writes, "were mailed not only in France but also . . . to collectors in England and throughout Europe."[66] To illustrate this, both Cary and Rainbird provide excerpts from the same advertisement, which Rainbird dates to 1902 and transcribes as follows:

WOMAN'S BELLY WITH ARTIFICIAL VAGINA

Giving to the man the complete illusion of life and procuring to him sensations as sweet and as voluptuous as those of the woman herself.

Externally this apparatus represents the belly with the exception of the legs. The secret parts, the mons veneris, are covered with an abundant and glossy fleece, the wings of the vulva, the nymphae and the clitoris offer themselves to the covetous glances as delightfully as the love-quim of the woman. Internally, the vagina possesses the enclosing folds which provoke the ejaculatory spasm. The contact is soft and the pressure may be regulated at will by a pneumatic tube.

On the other hand, a lubricating apparatus filled beforehand with a lukewarm and unctuous liquid sheds under pressure its contents between the vaginal inner sides in the same way as the feminine glands pour themselves out at the psychological moment.

It can be filled with air and emptied at will and be hid in the pocket as easily as a handkerchief or other toilet object.

The complete apparatus: 100fr.[67]

There are some elements of this advertisement that should give us pause. References to this ad, through Rainbird and Cary, have often been used in more contemporary work as evidence that early sex dolls were truly full-body, elaborate contraptions. Yet what is actually described here is just the "belly" (*ventre* being French slang for a vagina) with no legs. Also, the notion that the apparatus on offer could be deflated, folded up, and tucked away in one's pocket certainly does not suggest a large or elaborate device. However, what stands out most in this advertisement is the grand promises it makes. This artificial vagina will, apparently, offer its user "sensations as sweet and as voluptuous" as penetrative sex with a human being. The replica clitoris on the item, we are told, is itself so realistic that it is just as "delightful" to gaze upon "as the love-quim of the woman." Even if it is just an "artificial belly," rather than a full-sized doll, the advertisement makes the device in question seem quite complex, with vaginal "enclosing folds," a pneumatic tube that one can use to regulate pressure, and a "lubricating apparatus" to replicate the sensation of vaginal fluids: liquids that the device oozes and "sheds under pressure." All in all, for the price of 100 francs, it does sound like the potential customer stands to purchase an impressively elaborate contraption.

The reality, however, is so starkly different it is nearly comical. In the eleventh hour of writing this book, my many attempts to track down the original advertisements that Rainbird and Cary quote from finally turned them up in a fragile, hodgepodge volume currently held at the British Library—part of a private collection of erotic material amassed by, of all people, George Mountbatten, second Marquess of Milford Haven, the uncle of Queen Elizabeth II.[68] This volume contains both the original advertisement that Rainbird reproduces in her text and a later version of the same ad, this one from 1908, published six years after the first. Crucially, the

1908 ad includes not just the same advertising copy as transcribed, but also a detailed illustration of the *ventre de femme* itself (figure 2.5). For the seller's grand promises of "sweet sensations" and "love-quims," the item is little more than a lima-bean-shaped inflatable rubber cushion fitted with an ever-so-slightly vulva-shaped pocket. Not only is it incredibly basic and entirely unrealistic, it lacks almost all of the features that the ad accompanying it blatantly proclaims it includes, such as an "abundant and glossy fleece," the "wings of the vulva," or a clitoris. Whether it can truly be filled with liquid that is released under pressure is unclear. As for that pneumatic tube that was supposed to allow for carefully calibrated pressure, it appears to be a simple nozzle for filling the item with air, like a valve for inflating a bicycle inner tube.

This image is remarkable (though it may seem decidedly unremarkable) for two reasons. First, across multiple years of research, it is the only visual representation I have found that depicts what items of this sort actually looked like during the period: suggestive of the contemporary silicone torso sex dolls of the twenty-first century, but far, far simpler in design (figure 2.6). Second, this image makes clear just how enormously fantastical the rhetoric of advertising that surrounded the sale of sex toys such as this

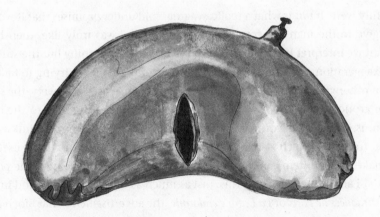

VENTRE DE FEMME

Figure 2.5
Artist's rendering of an inflatable rubber *ventre de femme*, advertised in a catalog from 1908. Based on an image from Album 7 of the Milford Haven Collection housed at the British Library. *Source:* Original art by Kathryn Brewster.

Torso sex doll (silicone)

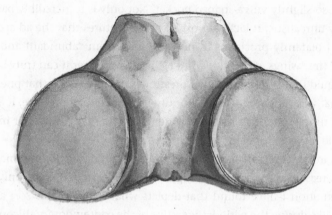

Figure 2.6
Artist's rendering of a contemporary silicone sex doll that replicates only the pelvic area. *Source:* Original art by Kathryn Brewster.

truly were. If *this* is what a replica vagina, sold under promises that it would "give to the man the complete illusion of life," was truly like, then how can we interpret the rhetoric found in such ads as anything but the stuff of exaggeration, euphemism, and the dulcet tones of a seller trying to talk up an otherwise unremarkable product? The language used is flowery, the stuff of erotica and bedroom talk, where genitals "offer themselves to the covetous glances." By contrast, the material reality of this item is clunky and far from "high-tech"—though, as I discuss in chapter 4, the rubber used to make such an item can itself be considered a sexual technology. The point that I am trying to make is this: Just as much as the erotic tales found in *Les Détraquées de Paris* or *La Femme endormie*, the advertisements that Bloch and others after him have mistaken for literal descriptions of sex dolls for sale at the start of the 1900s are in fact themselves wildly fictional. Far more than they sell actual devices through reference to their specific material properties, they sell a fantasy.

Admittedly, this hyperbolic tone is not unique to advertisements for artificial vaginas or sex dolls. It predominates in the broader genre of French ads for "rubber goods" and related items across the 1890s and the first decades of the 1900s. For instance, in an 1891 edition of the newspaper *La Lanterne*, Parisian rubber manufacturer Maison A. Claverie ran an advertisement promising the "only truly unbreakable condoms," which it described as "the last word in sturdiness, convenience, and safety."[69] Those interested in purchasing condoms from Claverie were instructed to send money to a local address to receive in return a "sealed box defying all indiscretion." Four years later, in an advertisement that ran in an 1895 issue of the newspaper *La Grisette*, we find that Maison A. Claverie has grown even more audacious in its advertising efforts. Now, in addition to condoms, it offers readers "special apparatuses for intimate usage (men and women)," available to view in an "elaborate illustrated catalog" that reportedly includes 220 engravings and six "lovely samples" (figure 2.7).[70] Positioned alongside a slew of other dubious newspaper ads for everything from "hygienic liquor" to pills that supposedly purify the blood of "impurities caused by syphilitic inoculation," the veracity of Claverie's claims seems far from certain. Rather, what the ad offers primarily is enticements and assurances: these condoms will not break; your privacy will not be violated; you will feel sexual pleasure when you receive our catalog.

Such rhetoric also pervades the contents of the rubber goods catalogs themselves, where the relationship between selling sex as a fantasy and selling sex as a technology becomes clear. One such catalog dated from 1900, this time from the Parisian seller Maison L. Bador, proclaims on its cover that its contents reflect the "perfected manufacture of inflatable rubber."[71] On the very first page, a note to clients insists, "All of our items are of the highest quality."[72] Many of the items presented in the catalog are accompanied by language that not only makes claims as to their superlative

Figure 2.7
An advertisement for Maison A. Claverie, a Parisian rubber goods manufacturer and seller of sex-related items, which appeared in *La Grisette* in 1895.

quality, but also specifically plays up their supposedly advanced technological characteristics. Take, for example, a particular condom given the rather
indulgent name *Le Bijou Velouté* (the velvety jewel; figure 2.8). From the
looks of the illustration, the condom seems unremarkable. Yet the catalog
text that accompanies the image reads: "Since 1890 when we founded our
company, many of our clients have asked for a condom that can withstand
the heat. After long and patient research based on our in-depth knowledge
of rubber, we have succeeded in producing an object of total perfection and
in the first material to leave the best known rubbers far behind."[73] In the
same catalog, Maison L. Bador advertises another condom under the even
more grandiose name *Le Roi des Dilatés* (the king of the inflatables).[74] This
"king" condom, we read, has been made "scientifically infallible." These
claims are rendered all the more comical when we consider that rubber
condoms from this time, far from being the thin sheaths we are accustomed
to today, were thick, cumbersome items—definitely not what we would
describe today as the stuff of velvet or kings.[75]

Here then lies the supposed historical basis for later descriptions of the
dames de voyage, in instances when they are imagined as elaborate predecessors to the sex robot: distorted, flowery marketing claims that offer
consumers a fantasy that seems to have been, in many instances, significantly far from reality. Through these ads, we observe yet again how fantasy
has come to shape narratives of sexual histories. And, once more, we see
through these examples how this fantasy was itself—long before the contemporary buzz around sex tech—already simultaneously about the allure
of sex and the allure of technology. The veneer of technological and scientific inventiveness that companies selling sexual devices used to promote
their goods suggests that part of what made these items alluring, already
more than a century ago, were their properties as pieces of technology.
With its promises of realism and unprecedented pleasure, the rhetoric of
selling sex tech sounded surprisingly similar in the year 1891 to how it
sounds today.

Fantasizing History, Reclaiming Fantasy

As we reach the turn of the twentieth century, what have we found in
our continuing search for the dames de voyage? The advertisements discussed in the previous section represent the earliest sources mentioned in

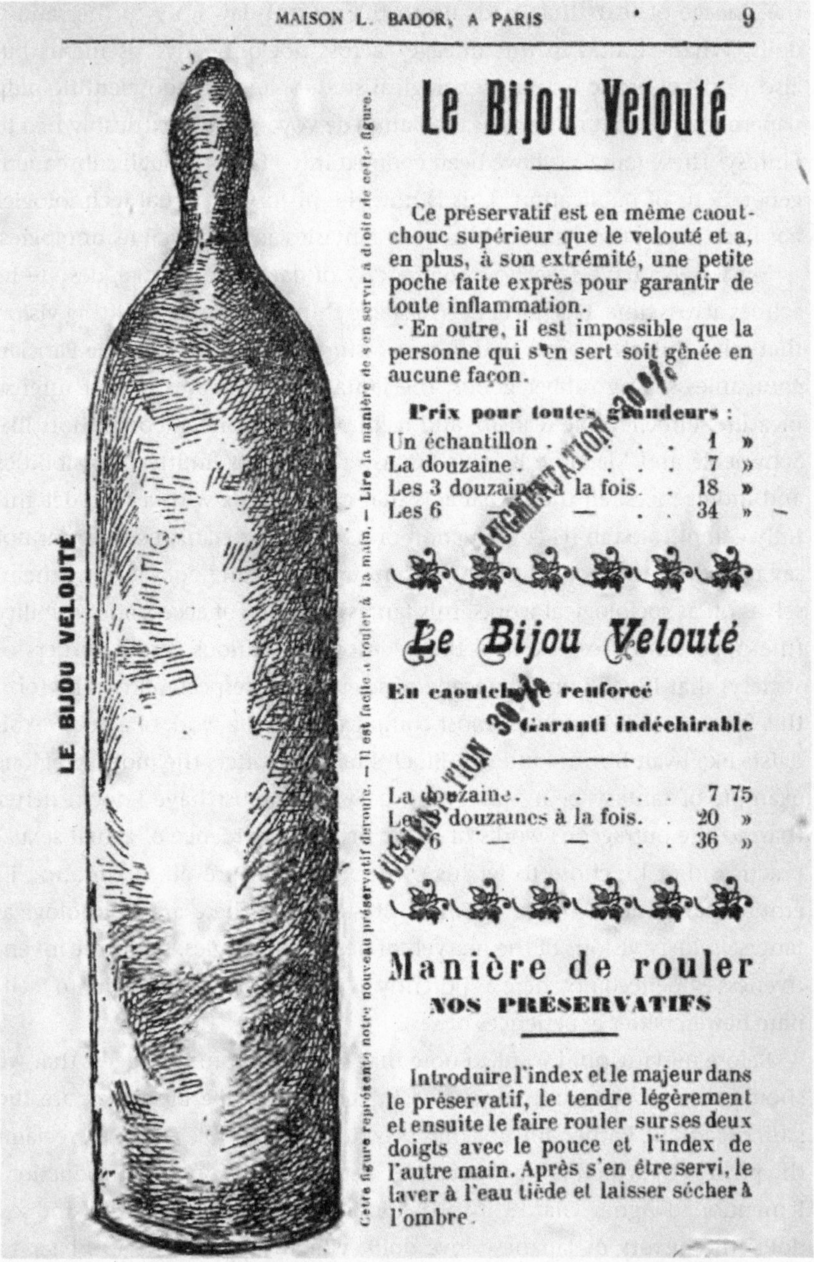

MAISON L. BADOR, A PARIS 9

LE BIJOU VELOUTÉ

Cette figure représente notre nouveau préservatif déroulé. — Il est facile à rouler à la main. — Lire la manière de s'en servir à droite de cette figure.

Le Bijou Velouté

Ce préservatif est en même caoutchouc supérieur que le velouté et a, en plus, à son extrémité, une petite poche faite exprès pour garantir de toute inflammation.

En outre, il est impossible que la personne qui s'en sert soit gênée en aucune façon.

Prix pour toutes grandeurs :

Un échantillon	1 »
La douzaine	7 »
Les 3 douzaines à la fois .	18 »
Les 6 — . .	34 »

AUGMENTATION 30 %

Le Bijou Velouté

En caoutchouc renforcé

Garanti indéchirable

La douzaine	7 75
Les 3 douzaines à la fois. .	20 »
Les 6 — — . .	36 »

AUGMENTATION 30 %

Manière de rouler

NOS PRÉSERVATIFS

Introduire l'index et le majeur dans le préservatif, le tendre légèrement et ensuite le faire rouler sur ses deux doigts avec le pouce et l'index de l'autre main. Après s'en être servi, le laver à l'eau tiède et laisser sécher à l'ombre.

Figure 2.8

A page from a 1900 catalog of rubber goods from Maison L. Bador advertising *Le Bijou Velouté* condom.

the lineage of texts that leads up to the present-day story of the sailors' dolls. What we find in this lineage—across not only advertisements but also works of erotic fiction, sexological studies, and pseudoscientific pulp paperbacks—is that notions of the dames de voyage are inextricably tied to fantasy. These fantasies have been codified into "fact" through subsequent generations of publication. This is how the history of sexual technologies has been made, through the history of fantasies about sexual technologies.

Yet we also have seen how the fantasy of dames de voyage, despite its echoes across time, has never been a stable thing but rather a shifting vision that takes different forms in different historical moments. For the Parisian companies selling rubber goods, the fantasy on offer was one of intense pleasure, unbelievable realism, and technological wonder. For authors like Schwaeblé and Madame B., this fantasy was one of limitless possibilities and indulgences: all the sexual acts that can be done with a doll, delightfully complicated in its construction or its inner mechanisms, who cannot say no. For the authors of those midcentury American books passing themselves off as sociological works, this fantasy was one of access and visibility (the opportunity to gaze into the bedrooms of various sordid corners of society) that hinged on the facade of science and respectability. The form this fantasy takes is perhaps most complicated in the work of actual sexologists like Iwan Bloch—and yet, Bloch's text also offers the most egregious example of fantasy being taken for fact. Bloch must have known better than to cite outrageous works of erotic fiction as evidence of actual sexual practices, but he chose to let his own excitement prevail, luxuriating in erotic imagination. Above all, the fantasies found here are technological fantasies, lusty visions of the marvels of new technologies, their true inventiveness evidenced nowhere as potently as in their imagined ability to facilitate newer, better experiences of sex.

Before moving on, I want to note that I do not mean to suggest that we should dismiss the works discussed in this chapter because they are the stuff of fantasy. Rather, my goal has been to render visible and even reclaim the process by which fantasy becomes history. In this book's introduction, I mentioned Agnès Giard's investigation of another origin story for sex dolls: the history of Japanese love dolls, which supposedly dates back to the 1680s.[76] Giard writes that contemporary accounts of this history, which are factually inaccurate and largely invented, are evidence of what she calls a *collective phantasmagoria*: a sort of shared, frenzied agreement to project the sexual fantasies of the present onto the past. The stories that surround the dames de voyage similarly reflect a collective phantasmagoria, one that operates in the present but also stretches back across history. That phantasmagoria merits critique, of course, but it also offers an intriguing framework for conceptualizing the heated work of making history. Cases like the dames de voyage or Japanese love dolls bring to the surface the sexual qualities of the multifaceted fantasies on which history is built. They show us how the telling of history, unmoored from fact, can become a kind of erotic daydream—or, maybe, how it has always been one.

This idea that history is built around a collection of shared fantasies is made nowhere more apparent than in the histories of sex and technology. Both hinge on romanticized visions of days gone by and fantasies of teleological "progress." Such fantasies do not remain merely whimsical, cultural, or even political, however. They are also literally erotic. The case of the dames de voyage and their appearance across twentieth-century texts demonstrates how the history of technology, though it may be presented as the stuff of bloodless innovation, has formed around lust and longing. What's more, the dames de voyage show us how the making of historical fact out of sexual fantasy itself has a history. If we follow stories of technology back far enough, we find ourselves face to face not just with the technological but also, inevitably, with the sexual.

3   The Birth of the *Dames de Voyage*: From Sex Workers to the Sexual Technologies of Sailors

If the story of the sailors' sex dolls as we read it today is a fiction, and the older documentation that supposedly supported the tale of the *dames de voyage* reveals itself to be the stuff of fantasy, where does that leave us in our search for the dames de voyage themselves? This chapter represents the third and final step in an attempt to locate the dolls as they are imagined in the present day: most commonly as rudimentary sailors' dolls made and used by European men during long voyages as sea, but also, at times, as early mechanized wonders that presage the twenty-first-century production of sex robots and other high-tech sex devices. The first two chapters of this book approached this search through paths of citational lineage and textual influence, moving backward from contemporary sources to see who does (and notably does not) reference whom and whether the breadcrumb trail of works that supposedly document the history of these dolls actually reflects historical fact. Ultimately, what the analyses presented in those two chapters revealed was more about how history is made through a variety of anxieties and fantasies. Neither attempt turned up solid proof of the existence of the sailors' dolls nor sufficiently explained what the dames de voyage truly were, if neither maritime traveling companions nor astonishing mechanical contraptions as they have been described.

Here, I take a different approach, turning away from established narratives about the dames de voyage and the history of sex tech. No longer retreading the path of existing scholarship, I look for the dames de voyage in new places. First, I do this by reconstructing a history of the term *dames de voyage* itself. The tale of the dames de voyage as it is told today is, we now know, a distorted construct of the story's transmission over the last 120 years. Yet the term *dames de voyage* is not the invention of scholars, pseudo-scientists, or pornographers. Rather, it predates its appearance in such texts, emerging from a set of vernacular uses and containing meanings that are not captured by the term's appearances in more official published works. Before Iwan Bloch took up the term *dames de voyage* in his 1907 sexological treatise that sparked generations of references to come, what did the term actually mean? And might this be the key to figuring out what the dames de voyage truly were? To find out, I delve into an archive of French quotidian newspapers and circulars, part of a Parisian print culture that thrived from the Belle Epoque to the start of the First World War, that document the term's vernacular uses.[1] This archive reveals the origins of a term that is itself central to many contemporary origin stories for sexual technologies, demonstrating how it shifted and eventually solidified in meaning through decades of everyday parlance.

In the latter sections of this chapter, I transition from a search for the origins of the term *dames de voyage* to a search for actual material traces of sailors' sex dolls. Even if the dames de voyage themselves were not sex dolls made and used at sea, this does not mean that such maritime creations did not exist. My investigation in this vein takes me through a very different set of archives, one that relates to—yet often fails to capture many aspects of—life in what we might broadly refer to as the *seafaring world*. I understand this world to encompass a set of histories, practices, and cultures that crossed Western Europe and North America (extending to the Pacific Islands, South America, Asia, and Africa through colonial occupation and trade) roughly between 1600 and 1900: a period that spans the many different moments in time in which contemporary authors claim the sailors' dolls were in use.[2] Here I look to sea chanteys, ships' cargo lists, and moralistic tracts about the sex lives of sailors, as well as to alternative sexual technologies of the seafaring world, such as figureheads and erotic scrimshaw.

What emerges in this last leg of the search for the dames de voyage—before I move, in the chapters to come, into the real origins of the sex doll and how we can make sense of the origin stories being told about sex tech—is a striking tangle of presence, absence, and possibility. Tracking down the shifting meanings of the term *dames de voyage* draws attention, ironically, to the people whom the term overwrites or even erases—most notably, the sex workers. Hunting through maritime archives also makes evident just how much of daily human life, and especially sexual life, such archives do not capture, whether they cannot or choose not to. And yet there is new potential here as well. By highlighting what has been overlooked or overridden, this work shows us what might have been and what could still be. It invites us to retrace the history of sexual technologies, recentering figures who have been crucial to the historical evolution of technologies and yet who are almost entirely rewritten out of those histories. By extension, it invites us to reimagine the origins of sex tech, looking not to the supposed creation of sex dolls at sea but rather to other, more revealing sexual technologies of the seafaring world as the place where the "high-tech" sex of today gets its start.

The Rise and Fall of the Term *Dames de Voyage*

The story of the emergence, evolution, and eventual decline of the term *dames de voyage* can be traced through the wealth of newspapers and other periodicals contained in the collection of the Bibliothèque Nationale de France and its archive, Gallica. These publications, which include everything from traditional news reports to satirical cartoons to lyrics for the newest songs, reflect a period in which inexpensive print materials sold in the streets of Paris (as well as elsewhere in France) were part of a vibrant and rapidly updating urban media culture.[3] As such, these documents capture the everyday events of (primarily) Parisian life, reflecting the shifting rhetoric of the day (figure 3.1).

References to dames de voyage first appear in these archives in the early 1890s. The earliest use comes from an 1893 short story, "En Des Jours pareils! Récit russe du temps de Pougatcheff" (In such days! A Russian story from the time of Pougatcheff), which was published in a quarterly digest of news and fiction from a French-speaking region of Switzerland.[4] The story, authored by M. le compte E. A. Sallins—likely one of the various grandiose pseudonyms found across the history of references to the dames de voyage—is a jumbled drama populated by military men and aristocrats on a far-flung journey. Some of these men are accompanied by women whom the author refers to as *dames de voyage* or *petites dames de voyage*: a flock of flirtatious companions who travel in their entourage.[5] These women characters are only mentioned in passing, but it is clear that they are meant to be living, breathing courtesans: sex workers acting as military camp followers, and certainly not sex dolls. They attend social functions like weddings and dinners alongside the male characters in the story with no particular comment from the author. The sexual nature of their presence, though not

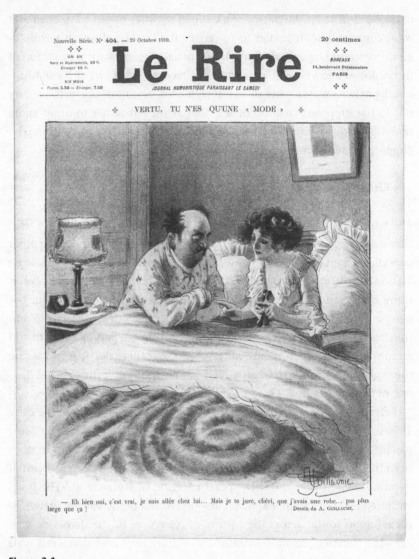

Figure 3.1
Cover of the October 29, 1910, issue of *Le Rire: Journal humoristique*, one of Paris's many quotidian newspapers and circulars.

made explicit, is reflected in Sallins's euphemistic, tongue-in-cheek use of the phrase *dames de voyage*, which he puts in quotations across its variations, suggesting a loosely veiled erotic meaning. These are "companions" for men on the move.

Indeed, other textual artifacts from this period suggest that *dames de voyage* likely referred to sex workers long before (and, for a stretch of years, at the same time as) the term referred to sex dolls.[6] This colloquial connection between sex work and women who "travel" is evidenced in one particularly potent site of erotic euphemism: the classified ads. In the final years of the nineteenth century and the start of the twentieth, it was apparently not uncommon to find messages among the *petites annonces* at the back of Parisian newspapers from men and women either looking for or offering themselves as "travel companions." This was a way of soliciting or advertising sexual services without making the sexual nature of the work in question explicit. We see this trend reflected in a newspaper article, in which the author jokes:

> We have all often read classified ads like this:
>
> Good, generous, loving man is looking for a travel companion. Excursions to Saint-Cloud, Constantinople, and Montmartre. Send photo.
>
> Young, blond, fun-loving woman would like to take small trips with a rich gentleman who has a car. Send offers to Mademoiselle Mercédès, 115, rue Pigalle, or come by between midnight and three in the morning.[7]

These are admittedly not real ads for traveling companions but rather comical amalgams representing the genre. A "loving gentleman" is looking for a friend to travel to such "exotic" destinations as Constantinople and Montmartre, the latter being a neighborhood of Paris notorious as a center of sex work during the period. As for the fun-loving Mademoiselle Mercédès, who herself lives on Rue Pigalle (another historical center of sex work in Paris), she is looking for a rich travel buddy who is invited to come visit her in the wee hours of the night. The joke in these faux ads hinges on just how obvious the euphemism has become: traveling together clearly stands in for engaging in sex. Going on "little trips"—*petites voyages*—is not about going anywhere at all, unless we count a trip to the red-light district.

For the term *dames de voyage*, the transition from meaning *sex worker* to meaning *sex doll* was not a seamless one. By 1903, fourteen years after the publication of "En Des Jours pareils!," sexologists like those discussed in chapter 2 began reporting that the term *dames de voyage* was being used for sex dolls that replicate all or parts of women's bodies.[8] Transcriptions of advertisements for inflatable rubber sex dolls dated as far back as 1902, like those seen in Henry Cary's and Evelyn Rainbird's books, also make use of the term.[9] Yet as late as 1910, evidence suggests that *dames de voyage* had not yet been widely adopted as a term for sex dolls in France, even when sex dolls themselves became the stuff of popular chatter. An article in the Parisian newspaper *Le Rire* (Laughter), published in a recurring section titled "Rire de la semaine" (Laugh of the week), recounts the story of a local woman who reportedly took her husband to court for a divorce.[10] Among the reasons that she gave for divorcing him: he had purchased a dame de voyage—that is, a sex doll. The article reads: "It was during a divorce lawsuit, recently pleaded, that this picturesque industry [of the sale of sex dolls] was revealed. The lady's lawyer told the court that the husband was cheating on her . . . with a woman made of rubber. Rubber? Absolutely. . . . The husband has defended himself by saying that he only bought one of these '*dames de voyage*' out of simple curiosity. 'I never touched her,' he declared. 'I swear!'"[11]

What makes this description of the dames de voyage striking, in part, is its incredulity. The article from *Le Rire* communicates a sense that a woman made of rubber, far from being an everyday item for sale, would be a bizarre and hilarious object. Even the husband in question defends himself by claiming he was simply curious about the novelty of such an item. As for the term *dames de voyage*, it is placed in quotes, suggesting once again its qualities as a term of art familiar only to those already inducted in the selling and purchasing of sex dolls. The article also replicates what it claims are the contents of the advertising flyer from which the husband learned about the rubber doll for sale (figure 3.2). It reads:

DAMES DE VOYAGE

Made of rubber
Full package, your choice of blonde or brunette.
(All models.)
Total price: 55 francs.
N.B.—with interior palpations: 75 francs.[12]

The author of the article raises some reasonable questions: "What are these interior palpations? . . . Can a husband surprised *in flagrante delicto* with a rubber woman be convicted of adultery?"[13] We might add: What does it

> # DAMES DE VOYAGE
> ### EN CAOUTCHOUC
> **Complètes, blondes ou brunes au choix.**
> (Tous les modèles.)
> *Prix net : 55 francs.*
> *N.-B.* — Avec palpitations intérieures : 75 francs.

Figure 3.2
Contents of an advertisement for dames de voyage, as replicated in an article in *Le Rire: Journal humoristique.*

mean for a sex doll to be the "full package" (*complètes*)? And who knew that long before contemporary, high-end sex manufacturers offered a range of looks and personalities for their dolls, a person could already order a rubber woman in blond or brunette? Newspaper reporting on this incident illustrates the disbelief that surrounded the concept of the dames de voyage as late as 1910, when the term, and indeed commercial sex dolls themselves, were still seen as humorous oddities.

It is not until the 1920s that we see *dames de voyage* used in colloquial French to mean, simply and without the need for clarification, sex dolls. Items previously referred to as *dames de voyage en caoutchouc* (traveling companions made of rubber) were by this time now simply referred to as *dames de voyage*, suggesting that they no longer need a qualifier to disambiguate inanimate sexual companions from human ones. In his 1929 novel *Êtes-Vous Fous?* (Are you crazy?), for example, René Crevel runs through a list of items on display in a feverishly imagined version of the Institut für Sexualwissenschaft (Institute of Sexology) in Berlin.[14] These include photos and drawings of all sorts of sexual activities, as well as an impressive collection of whips, chains, large phalluses, and "a lovely variety of *dames de voyage.*"[15] By this point, as Crevel's list illustrates, *dames de voyage* has entered a pantheon of various terms for sex toys and related items, no longer meriting explanation.

The last gasps of vernacular French references to dames de voyage crop up in the 1970s. In Roger Peyrefitte's 1973 novel *Des Français* (French people), a character muses on the state of consumer air travel (a very different mode of *voyage* than the trips suggested by the personal ads seventy years before).[16] Says the character, today "customs officers have only to reach their hands in suitcases to find themselves pulling out dildos and *'dames de voyage.'*"[17] As in Crevel's novel, Peyrefitte seems to feel that his use of the term *dames de voyage* needs no clarification. It clearly refers to a sex toy, not a person—and, moreover, to a sex toy small or flat enough that it can be stored unassumingly in a suitcase, calling to mind earlier images of replica vaginas that could indeed be described as "travel-sized" (figure 3.3). This tableau of the customs official reaching into a piece of luggage and pulling out a dame de voyage also illustrates how the concept of travel associated with the "traveling companion" has been variously repurposed with new meanings over time. What makes the term *dames de voyage* so well-suited as a euphemism for sex dolls is precisely that the dolls themselves can be packed up, stowed away, and traveled with. Especially if we consider that, as this quote suggests, the items referred to as *dames de voyage* were likely deflatable replica vaginas and not actually large-scale sex dolls, we can see how the *voyage* in *dames de voyage* comes to stand in for the items' roles as highly portable lovers.

By the end of the 1970s, we find that the term *dames de voyage* has acquired the air of a relic, becoming a somewhat distasteful turn of phrase. It pops up, for instance, in Vladimir Volkoff's 1979 spy novel *Le Retournement* (*The Turn-Around*).[18] In one scene, a woman anxiously insists to her lover, "I'm not interested in being one of these dolls to be tortured, one of these *dames de voyage*."[19] While *dames de voyage* does mean sex doll here, its connotations are far from neutral. Instead, it is spat out with disdain, connoting a mere object to be abused and discarded—further evidence that the term, even when it has been used to refer to sex dolls, has not been associated with the mechanical wonders described by sexologists and historians across the twentieth and twenty-first centuries. Rather, as it appears here, *dames de voyage* signifies something cheap, dirty, and perhaps alarming: no longer the stuff of curiosity, but instead the stuff of questionable relics on dusty shelves at the back of sex toy shops.

This trajectory, from the emergence of the term in the 1890s to its relative disuse by the 1970s, paints a picture of the rise and fall of the dames de voyage—not as figures in accounts of sexual histories but rather as items that circulated, in large part, through discourse. While this presents an image of the dames de voyage that is far less grand than other historical

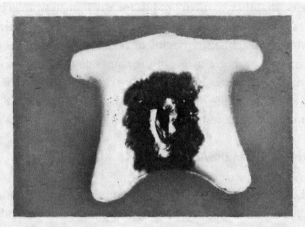

Vaginal-Ersatz
Onaniebehelf eines im Institut für Sexualwissenschaft in therapeutischer Behandlung gewesenen Patienten, von diesem eingesandt
(Archiv des Instituts für Sexualwissenschaft, Dr. Magnus Hirschfeld-Stiftung, Berlin)

Tab. V   Zur Sittengeschichte der Onanie

Figure 3.3
Small replica vagina labeled as having been used in a "therapeutic treatment" for a patient at the Institute of Sexology in Berlin. *Source:* Leo Schidrowitz, *Ergänzungswerk zur Sittengeschichte des Lasters* (Supplement to the moral history of vice) (Vienna: Verlag für Kulturforschung, 1927).

accounts have suggested, it nonetheless offers useful insights. Whether or not European sailors had sex dolls as early as the 1600s, the first dates mentioned by contemporary writers who tell the tale of the dames de voyage today, the term could certainly not have been used as a name for these dolls during that early period, given that it did not surface until the end of the 1800s. Indeed, as we have seen, the term wasn't even commonly used to refer to sex dolls until the early 1900s.

In addition, the history of the term *dames de voyage* helps bring to the surface the unacknowledged place of sex work in the origins of sex tech—or, at least, in sex tech's imagined origins. As a euphemism, the term *dames de voyage* itself both does and does not acknowledge sex workers. Historically, once the term comes to mean *sex doll*, the role of actual sex workers in the term's history becomes doubly obfuscated. Yet the history of the term nonetheless imbues the dolls themselves with an association with sex work: a sense that they stand in not just for any human women but specifically for escorts. Even as contemporary histories of sex tech attempt (often but not always) to frame the origins of sexual technologies as the result of men's inventive labor, the tale of the dames de voyage brings with it a reminder that sex workers have long played a crucial role in the formulation, use, and cultural life of sexual technologies. In a very real way, it is *their* labor on which sex tech has been built.

Imagining Sex Dolls in the Arms of Sailors

Tracing the history of the term *dames de voyage* also reveals another key finding: the phenomenon of associating sex dolls with sailors, independent of sailors' actual use of sex dolls, itself has a history that can be tracked through these quotidian texts. The contemporary tale of the dames de voyage proposes that sailors and sex dolls are a natural match, with the conditions of men's lives at sea supposedly sparking creative solutions designed to satisfy their frustrated desires. It turns out that this line of thinking is not just a thing of the present. It also manifests, in forms that are related yet different, in the past.

Associations between sex dolls and sailors begin to make appearances in this archive as early as 1885 and continue through the early decades of the 1900s. We see examples in works of fiction like the erotic tale *Le Femme endormie* (The sleeping woman), discussed in chapter 2, which includes an aside claiming that rubber sex dolls were "invented for the pleasures of Captain Pamphile," a reference to the seafaring protagonist of Alexander Dumas's 1839 novel, which I return to in chapter 6.[20] Another particularly intriguing example is a story that ran in the circular *La Grisette* in 1895 titled "La Femme du capitaine" (The captain's wife).[21] This comic tale features a lusty sea captain named Falot, who bemoans the lack of women aboard his ship and longs to feed his sexual appetite. The ship's doctor proposes a solution. He asks the captain for 200 francs for a mysterious purchase, assures him that the item he is about to acquire is just as good as the "real thing," and then heads off on a mission as soon as the boat reaches the port of Le Havre.[22] We read:

> And so [the doctor] took the next train to Paris. As soon as he arrived, he went to the Palais-Royal area, to a little nearby street.

Once his shopping was done, he got back on the express train, with his purchase carefully packed up in a coquettish box of waxed walnut. It was the size of an ordinary suitcase, complete with a metal handle.

Anyone who saw him would have taken him for a traveling photographer, but he was not worried about taking in the view. His only concern was that someone would catch a glimpse of the contents of the thing he was carrying, which could have accidentally opened up in the middle of the road.

As soon as he returned, he quickly went aboard the ship and immediately headed to the captain's cabin. . . . "Here," said the doctor, as he placed the fake suitcase on the table.

Falot's heart immediately started racing. He was full of joy and anxiety as the doctor proceeded to open the box . . . It was nothing other than a superb woman in rubber. Nothing about her was lacking, you hear me, nothing![23]

Pleased with this new rubber companion—whom the captain and the doctor quickly pull out of the suitcase, assemble, and fill with warm water for a "most appealing effect"—Falot heads off once again to sea, this time in the company of his new "wife," the doll.[24] The story ends when his real, flesh-and-blood wife (whom he was apparently married to all along) questions him about the mistresses he has surely taken during his journeys. When she learns the truth, she embraces the doll and thanks it for reducing the number of illegitimate children her husband would have otherwise produced.

"La Femme du capitaine" is a farcical story, but it is also revealing. The description it offers of a *femme en caoutchouc* (rubber woman), who is sold in pieces and comes packaged in a "coquettish box" resembling a suitcase, is actually one of the more detailed surviving accounts of what actual rubber dolls sold at the time may well have actually been like.[25] More relevant to our present purpose, the story demonstrates how the association between sex dolls and sailors, rather than simply being a trait of contemporary accounts, also had a place in the cultural imaginary around sex dolls as far back as the 1880s and 1890s. However, there are also notable differences between this earlier vision of sex dolls at sea and those commonly printed in histories of sex tech today. Here, the sex doll is seen as an expensive item only available to a ship's captain, not its crew. Falot does not cobble together his sexual companion from scraps of cloth or leather; he gives an employee a large amount of money and receives a doll constructed by professionals. Thus, though we see echoes of the associations between sex dolls and sailors across time periods, what forms those associations take remain distinct. In "La Femme du capitaine," the sex doll might be funny, but the joke is really on the captain. He is not portrayed as inventive and hearty in his manly appetites. Instead, he is portrayed as an oversexed boor with money to blow on a ridiculously elaborate portable wife.

Other examples demonstrate how associations between sex dolls and travel extended beyond sailors per se and more generally to European travelers abroad. As late as 1921, a story appeared in the newspaper *La Vie Parisienne* (Parisian life) called "La Poupée" (The doll).[26] Opening with a fitting reference to the tales of E. T. A. Hoffman, whose automaton character Olympia always looms in the background of Western fantasies about mechanized lovers, "La Poupée" is a bawdy tale—this one less comical and more menacing than "La Femme du capitaine."[27] It tells the story of a French colonial "explorer" who has recently returned from an expedition to Sudan. One night, after a trip to the opera, he brings a friend back to his home to introduce him to his lover, Mademoiselle Ketty. Ketty, who looks like a dewy-eyed teenager and who cries out for her "mama" when squeezed, turns out to be an elaborate painted rubber doll. The explorer, who has undressed the doll and positioned her sitting on his knees, describes how he purchased her in London at an outfitting store for 6,000 francs. "She was my mistress for two years, over there, in Africa," he explains. "I must say that with Ketty, there's no drama. . . . It's charming, to have a woman who is not an imposition, who is not jealous, who never gets a migraine, who does not have whims, and who does not speak."[28]

In this scene, the rubber sex doll—who is pale-skinned, rosy-cheeked, and silent—is presented as the opposite of those women who are deemed less desirable: women in Africa (whose implied Blackness contrasts starkly with Ketty's unnaturally white skin), women who have physical or emotional experiences of their own, and women with voices. This sheds light on a racialized logic, which persists to this day, that sex dolls are particularly well suited to white male travelers because the actual women they encounter on their journeys are likely to be unacceptable—in a hegemonic and deeply problematic sense—as sexual partners. "La Poupée" is also positioned among a series of illustrations designed to be comical and titillating, including one full-page image of disembodied women's legs: a fetishistic scene of slender fleet, pointed high heels, and stockings arrayed for view. The caption beneath the image reads: "Nature gave women legs so that they could make men walk."[29] Viewed alongside the story preceding it, the image captures the spirit of "La Poupée," which presents women's bodies as objects and suggests that those bodies have meaning, first and foremost, when they spur a man on in his travels.

Contemporary accounts insist that there is a natural link between sex dolls and sailors precisely because (straight) men's desires are themselves imagined to be timeless. Yet as the examples presented here demonstrate, the association between sex dolls and sailors has itself been the product of specific historical and cultural contexts. Consider two overlapping sets of factors that likely together sparked this earlier imagined link between sailors and sex dolls: one a matter of changes in commerce and the other a matter of changes in maritime norms.

First, the decades in which we see stories emerge in Parisian newspapers linking sex dolls and sailors are also those in which commercial sex dolls were themselves being advertised for sale by Parisian rubber goods sellers.[30] At least one source attests that these advertisements explicitly promoted the dolls as recommended "for the use of naval officers and others who are deprived of female society for long periods of time."[31] Thus, the connection between sex dolls and sailors may well have been in part a construction of businesspeople who made and sold such dolls. It is questionable at best whether ads of this sort were truly designed to persuade naval officers to buy contraband inflatable sex dolls. Far more likely, this reference to sailors served a function very similar to the function it serves in contemporary accounts of the dames de voyage today: to legitimize sex dolls by portraying them as practical items for upstanding maritime citizens who simply happen to be "deprived of female society."

Simultaneously, a second factor may have influenced the rise of the cultural imaginary that linked sex dolls with sailors: shifting historical practices around actual women's experiences at sea. Many of the contemporary accounts of the tale of the dames de voyage are premised on the assumption that sailing ships during the period were always all-male environments. Maritime studies scholars have shown this to be a myth. In addition to the many women involved in the seafaring world through their roles at home or at port, there were historical examples of female sailors and pirates—as well as women who came aboard in their capacity as family members of the crew.[32] Haskell Springer, in an essay titled "The Captain's Wife at Sea," explains how some wives of captains accompanied their husbands on their voyages, "a practice that appears to have begun in the late eighteenth century and peaked in the late nineteenth or early twentieth century" in

America and Europe.[33] This is a telling overlap: in the same historical period when ships were most likely to sail with actual captain's wives onboard, we see the emergence of stories about captains taking manufactured wives (i.e., sex dolls) with them to sea.

Thus, it seems that the association between sex dolls and sailors cannot be credited to some fundamental male desire mixed with masculine ingenuity. Instead, it is the result of factors related to mass media and governance: the particular platform of the urban newspaper, the constraints placed on publicly advertising sex, and the actual role of women like sex workers and captains' wives in the histories at hand. Thus, both the term *dames de voyage* and the association between sex dolls and sailors were not created in the absence of women (as the vision of men "deprived of female company" at sea would suggest) but rather in cultural negotiations of their presence.

Absence in the Archives: Tracking Traces of Sex in the Seafaring World

This research indicates that the term *dames de voyage* never historically referred specifically to sailors' dolls, and indeed that the association between sailors and sex dolls was an imagined construct as much in earlier periods as it is today. However, that does not preclude the possibility that Western sailors somewhere between the seventeenth and the twentieth centuries really did make and travel with sex dolls. If not to the citational trail of the tale of the dames de voyage (as in chapters 1 and 2) or to the actual use of the term (as in the preceding sections), where might we turn for one last effort to find the sailors' sex dolls?

One obvious place is the field of maritime studies. Unfortunately, existing work on histories of gender and sexuality in seafaring offer no accounts of sex dolls used by sailors. However, this research does provide helpful historical context for such a search. For instance, one thread in maritime studies scholarship addresses the experiences of women across the history of the Western sailing world (with an implicit focus on white European and American women).[34] These works often strive to surface the women who contributed to seafaring history—whether by participating in the social structures that surrounded maritime life or by dressing as men and joining crews—but who have been largely omitted from historical narratives.[35] Although such work focuses on actual women rather than dolls, and though its engagement with sexuality is minimal, it does helpfully challenge the simplistic vision of the sailing world as one made up entirely of men: a vision that itself animates the interwoven sexual and technological fantasies underlying the contemporary tale of the dames de voyage.

Another thread within maritime studies and adjacent fields explores the sexual behaviors of sailors, though work in this area too makes no mention of sex with dolls.[36] This work largely discusses either sailors' engagement with prostitutes in port towns or same-sex contact between men at sea. For example, in her book *Slavery at Sea: Terror, Sex, and Sickness in the Middle Passage*, Sowande' M. Mustakeem explains the close relationship between sailors during the period of the Atlantic slave trade and economies of sex on shore:

> Moving across the Atlantic traveling from port to port, life on the waterways granted mariners access to highly sexual behaviors and opportunities. Familiarity and duration of time bore on their participation in an economy fueled by negotiations for and allocation of sexual services. No matter their personal interest, the dominance of brothels and establishment of personal businesses for the hiring of prostitutes—enslaved, free, white, and black—sailors were exposed to an intimate world created in port communities deeply entrenched in the buying and selling of sex.[37]

This "intimate world" of the port communities that Mustakeem describes forms a backdrop to both the real and imagined histories of sexuality in the seafaring world. Indeed, in part, stories like the tale of the dames de voyage can be seen as attempts to distance the history of sailing (and, by extension, the history of sexual technologies) from the history of sex work—a kind of disavowal that echoes the erasure of sex workers from the term *dames de voyage* itself.

Sex between men aboard sailing ships is similarly a sensitive subject in maritime histories. While writing on seafaring and prostitution addresses the sex lives of sailors in port, writing on sex between sailors considers what sexual practices they may have performed at sea. Some of this work takes the idea of homosexual contact between sailors literally, searching for evidence to support a more contemporary expectation that men aboard sailing ships surely slept with other men.[38] However, this hunch has proven hard for historians to corroborate given that sodomy was illegal in many of the countries under whose flags these ships set sail—such as the British Royal Navy, which counted sex between men an "unnatural offense to be punished by death."[39] (Some historians have focused on the prospect of same-sex contact between men who stood outside the law—namely, queer pirates—a subject I address at greater length in chapter 5.) Surviving records center on a handful of court martial cases in which men were charged with the crime of sodomy.[40] A greater proportion of scholars have focused on the queer implications of seafaring literature, drawing out homoerotic themes in fictional texts like those by Herman Melville.[41] Such work does not claim to offer actual historical facts about the sex lives of sailors, which might themselves help corroborate or debunk the notion that such sailors made and used sex dolls, but it does offer a productive counternarrative to the tale of the dames de voyage, centering men's bodies rather than women's as objects of desire.

Pushing forward in the search for actual sailors' sex dolls requires moving beyond dominant accounts of maritime history, however. It also requires diving into archives that document the everyday sexual cultures of the seafaring world. Yet hunting down traces of actual sailors' sex dolls comes with specific challenges. Such dolls, if they existed, are unlikely to have been mentioned in the forms of documentation most widely used by maritime historians, which include letters that sailors sent home to their families, professional communications, and published narrative accounts of mariners' voyages.[42] Documents in all of these genres would have represented life at sea as impressive and upstanding, not as the realm of sexual impropriety, whether in the quarters of a captain or in the crowded bunks of a ship's crew. In addition, contemporary accounts insist that the sex dolls in question were constructed out of scrap materials and shared among sailors, which suggests that they were made and used by lower-class seamen, whereas surviving written accounts largely reflect the experiences of ships' officers or doctors. Ships' cargo lists, which record many of the items aboard a vessel, proved to be another unhelpful avenue in my search. Standard practice, as I learned, dictated that such lists did not contain the personal belongings of crewmen.[43] Then there is the problem of slang. Based on the materials presented in the first half of this chapter, we now know that the term *dames de voyage* did not actually refer to sailors' sex dolls. So what would such dolls have been called? It's likely they would have been referred to using colloquial, veiled language: some set of keywords that might be crucial for navigating present-day archives and yet seems to have been lost to history—if it ever existed at all.[44]

Another place to look for the possible existence of sailors' sex dolls is historical sea chanteys. Even if sailors did not leave written records of their use of such dolls, perhaps the dolls found their way into sailors' songs. Indeed, in contrast to official documentation of life aboard ships, sea chanteys did often feature themes related to the daily lives of lower-rank sailors. The San Francisco Maritime National Historical Park offers the following background on chanteys in an info sheet for those interested in participating in the park's monthly group chantey sing:[45]

> In merchant vessels of the mid-nineteenth century, the tasks of weighing anchor, hauling on the halyards to raise sail, pumping ship, handling cargo, etc. were done by hand without engines or other forms of mechanical power. During this period a vast number of work songs were developed and used aboard ships as a way of rhythmically working in unison and allaying the boredom of this sometimes long and tedious work. . . . As they were passed from ship to ship, from shore to ship, and from generation to generation, each chanteyman put his own stamp on these songs, adding or discarding lyrics, varying the melodies, etc. It has been said that no chanteyman sung one the same way twice. International in nature, chanteys were made up and sung by Anglo and African Americans, Europeans, West Indians, African oarsmen, Italian and Japanese fishermen . . . in any part of the world where groups sang together to help get seagoing, cargo loading and fishing tasks done.[46]

This description of sea chanteys initially sounds promising. Because the creation of such songs was a widespread art form, reflecting the experiences of not only white European or American sailors but also those from Africa, the Caribbean, and Asia, as well as African Americans, we might expect sea chanteys to capture a particularly broad and diverse array of sailors' practices. Even the ways that chanteys were used—to help workers perform tasks in unison before the introduction of engines—makes the songs surprisingly relevant to histories of technologies. Such songs were themselves, it seems, a kind of premechanical folk technology that put the bodies of sailors into synchronized motion.

Unfortunately, in our correspondences, multiple experts on sea chanteys and other elements of sailing cultures confirmed that they had never encountered any references to sex dolls or other sex toys in such songs or other documents.[47] (Because many sea chanteys are only preserved through oral traditions or untranscribed early recordings, these experts themselves represent the most comprehensive sources of information about the contents of sea chanteys, rather than traditional print archives.) Scholar Gibb

Schreffler also helpfully explained that chanteys "by and large come out of an African-American vernacular and American popular music culture in which, in my opinion, there doesn't tend to be much interest in . . . bawdiness."[48] Thus, even in a genre of expression that captured the everyday lives of sailors, a subject like sex dolls would have been off-limits.

One last possible route for finding traces of sailors' dolls is accounts written about sailors' sexual practices during this period. These largely take the form of sermons and moralistic tracts, composed by community leaders in port towns, that denounce the behaviors of sailors who drink, gamble, and visit local brothels as dishonorable, dangerous, or simply seedy.[49] For example, an 1824 article that appeared in a publication called *The Christian Herald and Seamen's Magazine* (a notably specific venue) bemoaned the difficulties of enacting religious salvation for sailors, who spent their days in port drunkenly cavorting in boarding houses, partaking in the "sin of lewdness," and singing "noisy, obscene songs."[50] Similarly, in a 1906 article titled "The Sailor and City Problems: Where the Real Peril of the Seaman Begins," a reverend from New York lamented that sailors brought unscrupulousness and peril to otherwise god-fearing port communities—evidence, apparently, that the fall of Sodom and Gomorrah did not effectively cleanse the world's cities of their "wickedness."[51] In this piece, the reverend goes on to explain that the behavior of the sailors—whom he describes as "essentially masculine, strong sexed, else they would never be sailors"—devolves into debauchery.[52] This was due, the reverend claimed, to their "susceptibility to feminine influences which is fostered by a sea life where the sanctifying influences of women is totally absent."[53]

In this characterization of sailors as lusty, manly, and sexually deprived, we encounter an early version of yet another recurring feature of the contemporary tale of the dames de voyage. Yet despite their willingness to call out sailors' various improprieties, these accounts too make no mention of sex dolls. They seem concerned only with the sexual activities of seamen on shore; as for what happens at sea, that is another matter entirely. In addition, as scholar Amy Parsons pointed out in our correspondence, references to items like sex dolls would have fallen under the category of "unspeakable crimes" for authors such as these.[54] Thus, even in writing specifically focused on detailing the "lewdness" of sailors, the use of a sex doll would have been deemed too obscene to mention. This is yet another example of how sexual practices that fall outside of societal norms are erased from recorded history. Because such practices are viewed as deviant and abhorrent (whether we are talking about the purported use of sex dolls or very real phenomena like same-sex romances and non-normative performances of gender identity), those involved faced not only derision, discrimination, legal oppression, and violence, but also erasure from history.

Ultimately, none of this archival searching turned up any trace of the historic existence of sailors' sex dolls. That is not to say that sex dolls have never been taken aboard a ship—a possibility that surely became increasingly likely by the mid-twentieth and on into the twenty-first centuries with the growing availability of mass-produced dolls. Yet it seems we can now definitively say this: the tale of the dames de voyage itself is a myth. Western seamen sailing between 1600 and 1900 were not actually engaged in any regular or widespread practice of making or using sex dolls during their journeys. The idea that men cobbled together sex dolls at sea, passing them from sailor to sailor, emerges purely from the twenty-first-century imagination.

The Sexual Technologies of Sailors: Figureheads and Scrimshaw as Alternative Origins

However, our time with the maritime world is not quite over. The historic sailors' sex dolls may be the stuff of myth, but the search for the dolls reveals that there are many other practices and handicrafts of the seafaring world that we could see, in their own way, as sexual technologies. These form a set of alternatives to the myth of the sailors' dolls: other, more real possible origins we could point to in reimagining the history of sex tech as it relates to the history of the sea.

One example is ships' figureheads. During my exploration of the digital archives of the British National Maritime Museum, where I was hunting through ships' cargo lists for signs of sex dolls, I was struck by the museum's special collection of figureheads. These date from roughly the 1700s to the early 1900s, a time span not dissimilar to the range of dates various authors attach to the tale of the dames de voyage. In popular culture today, the notion of the figurehead conjures up a vision of a woman's body, crafted from wood, bursting forth from the prow of a tall vessel as it breaks dramatically through the waves. In reality, the figureheads in the British National Maritime Museum's collection are a more eclectic bunch. Some are animals, some are coats of arms, and some are women, depicted either as three-quarter-length figures or as busts.[55] Such figures vary in their appearance from the banal to the grand to the erotic. In their own way, figureheads fill a related symbolic role as the imagined sailors' sex dolls. They are, as maritime studies scholars have written, the "wooden women" (constructed objects) who accompanied the "iron men" (sailors who embody rugged masculinity) on their voyages.[56]

In particular, one figurehead from the British National Maritime Museum's collection stands out as a particularly evocative example of the genre (figure 3.4). Believed to have been mounted to a French merchant ship that was wrecked off the coast of Cornwall in the 1800s, the effigy has been crafted in the form of a blushing woman holding a small bunch of flowers.[57] Wearing a corseted dress decorated with blue and white flounces, her hair tightly coiffed and curled up in a style popular in the court of King Louis XV, she gazes out to sea with a placid expression and an almost sensual glow. Her lips are rouged and her breasts heave from the top of her garment, leading the way into the open ocean. The gentle positioning of her hands and the delicate closed fan held at her waist suggest a life of leisure. She is beautiful, refined, regal. Yet as a viewer of the collection regards the figurehead, their eyes cannot help but be drawn downward. Delicately crafted and carefully restored, this figure is affixed to a rough wooden beam, worn by time and the ocean. The beam is positioned coarsely between her legs. From the hips down, her body seems to disintegrate. Her paint thins. The white of her skirts yellows. Her delicate silhouette collides with the phallic remnants of the prow on which she now rests.

Although she is no sex doll, this figurehead cannot help but recall the dames de voyage. She too has been crafted to resemble a woman; she too would have traveled on long journeys, possibly the lone female figure amid a crew of men. Coming from France and ending up in the remnants of a crash off the shores of England, she also moved across European national borders. And though she is among the seemingly best-preserved examples in the British National Maritime Museum's figurehead collection, she comes from a shipwreck, suggesting that she, like the vision of the sailors' sex dolls thrown overboard found in some contemporary accounts, has spent time at the bottom of the sea. Her shining, rosy skin, visible chest, and cinched figure all give her a lifelike sexual allure. Both part of the ship and something other than it, she is simultaneously less than and more than human.

Figure 3.4
Circa 1800s figurehead from a French merchant ship wrecked off the coast of Cornwall. *Source:* National Maritime Museum, Greenwich, London (Creative Commons).

Meanwhile, the presence of the decaying wooden prow on which she is perched reminds us that she is a battered carving: an object after all. The jutting position of the wood, which reaches just to her pubic area, visually echoes an act of penetration. Yet despite all that she has been through, it is she—not her ship, not its crew—that survives, preserved as a representative of history. The figurehead is the sister to the dames de voyage. Both are dolls, of a sort, and both act as embodiments and instruments of desire at sea: the desire for sex, the desire for safety and good fortune.

Another seafaring craft that came to the fore through my search for sailors' sex dolls is scrimshaw. Scrimshaw is the art of making engravings on bones and other hard animal by-products, such as whales' teeth and walruses' tusks.[58] Historically, scrimshaw is most closely associated with whaling ships, on which some sailors spent their considerable free time intricately etching figures and scenes onto whalebone. Culturally, scrimshaw is often held in a certain reverence as a folk craft, and there are scrimshaw collections at numerous museums in both America and England.[59] Today, scrimshaw calls up visions of craftsmanship driven by the wearying circumstances of life at sea—a more quaint, staid version of the image of the sailor whose sexual frustration drives him to invent the sex doll. Gender and desire too frequently appear as subjects in scrimshaw, mostly in the form of renderings of women left behind: wives, lovers, mothers, and sisters whom the sailor longs for and therefore slowly, carefully recreates through art (an actual woman in his pocket, to point back to the advertisement mentioned in chapter 2 that promised inflatable lovers that could fold up as small as a handkerchief; see figure 3.5).[60] In this sense, scrimshaw can be seen as a handicraft done by men who have channeled their longing for women into more "appropriate" acts of creation, producing objects that are careful and loving, making manifest their loneliness and their ability to wait.

The case of a brief twenty-first-century controversy helps surface scrimshaw's surprisingly sexual implications. In July 2013, the Vancouver Maritime Museum hosted an exhibit called "Tattoos and Scrimshaw: The Art of the Sailor," which included a display case featuring nine works of pornographic scrimshaw. A flurry of news stories and online blog posts appeared in response, sparked when a woman who had visited the museum contacted a local news source to sound the alarm about what she referred to as "whalebone porn."[61] As for the items themselves, they depict a range of

Figure 3.5
Scrimshaw on whale's tooth: an image of a woman waiting for a sailor to return from sea. *Source:* Auckland Museum (Creative Commons).

scenes, from the more traditionally erotic to the wonderfully absurd. One features a tableau in which a sailor has sex his first night ashore. Another presents a trompe l'oeil peephole that reveals a masturbating nun. On a third, a jauntily positioned woman with her clothes falling off rides a giant penis, the testicles of which have been reimagined as bicycle wheels (figure 3.6).[62] These are not the whimsical scrimshaw scenes typically found in the museums and gift shops of New England. Perhaps unsurprisingly, the historical authenticity of this scrimshaw is apparently up for debate. News reporting about the exhibit cites experts who claim that the pieces are "too sexy" to derive, as others have assumed, from the nineteenth century.[63] They may instead be the work of a Los Angeles–based artist who has been making and selling "fake erotic scrimshaw" to well-paying collectors since the 1970s.[64]

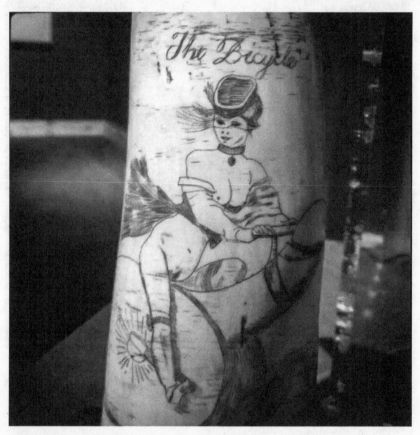

Figure 3.6
Pornographic scrimshaw displayed as part of "Tattoos and Scrimshaw: The Art of the Sailor" exhibit at the Vancouver Maritime Museum in 2013. *Source:* Sara Bynoe.

Whether these items are historically "real" is not the point, however. What matters is that they literalize and thereby make visible the underlying erotics of a men's maritime craft like scrimshaw. This "whalebone porn" proved scandalous because it was so unexpected (who painstakingly carves a giant penis bicycle onto animal bone?) and yet simultaneously so fitting (of course there is pornographic scrimshaw!). As a result, it casts a tellingly lusty shadow over the broader practices and products of an otherwise seemingly straitlaced historical tradition. Thus the making of scrimshaw, like the carving of figureheads, can itself be seen as an example of an erotic technology—at least, just as much as the imagined construction of a sex doll fabricated at sea.

Viewing figureheads and scrimshaw as sexual technologies challenges us to imagine creative new ways to talk about the origins of contemporary sex tech. What possibilities emerge if we jettison the tale of the dames de voyage, tossing it overboard and replacing it with another anecdote about the erotic lives of sailors? To say that sex tech started with figureheads, for example, would offer arguably a more feminist symbol for the sex doll's origins: a woman who leads the way at the prow of a ship, who can be seen but not touched, unless a sailor leans perilously close to the waves. Alternatively, to say that sex tech started with scrimshaw would bring an emphasis on artistry and a feminine-coded association with handicrafts to the history of sexual technologies. It would also draw our attention to links between the interrelated histories of sex and technology and histories of colonialism, highlighted through the use of materials like ivory collected through colonial exploitation and violence.

Although figureheads and scrimshaw are real material elements of seafaring history, they would be no more "true" as the subject of sex tech's origin stories than the dames de voyage. What these examples illustrate is the constructedness of origin stories themselves—as well as the infinite possibilities for telling histories of sex and histories of technology in new ways. The vernacular terms, cultural imaginaries, maritime crafts, and even archival absences discussed in this chapter all point toward the importance of moving beyond the tale of the dames de voyage, finding alternate histories and counterhistories for sexual technologies and "the very first sex doll."

4 "All Is Rubber!": The *Femmes en Caoutchouc* and the Actual Origins of the Commercial Sex Doll

As the myth of the sailors' sex dolls blows away like a fog cleared by an ocean breeze, another figure appears: not women of travel but women of rubber. The *dames de voyage* as they are envisioned in the present day may not have been real, as I have demonstrated over the last three chapters. But in searching for them, what comes to light is the actual history of early commercial sex dolls in Europe and America. This is a history that dates back not to the 1600s, as some accounts of the sailors' dolls suggest, but rather to the mid-1800s, when a different sort of "very first sex doll" was being dreamed of and then manufactured in Paris: dolls that have been, to date, almost entirely left out of histories of sex toys and sexual technologies. These are the *femmes en caoutchouc*, the "rubber women" or "rubber wives." If they sound familiar, it's because we have seen them briefly before, in advertisements for the dames de voyage and related news stories, like the 1910 account of a woman who divorced her husband partly on the grounds that he cheated on her with a dame de voyage made of rubber. Yet the story of the femmes en caoutchouc goes beyond their relationship to the tale of the dames de voyage. The rubber women come with their own complex cultural history: an alternative narrative that roots the origins of sex tech not in the ingenuity of straight male desire, as in the tale of the sailors' dolls, but rather in the shifting material industries of the nineteenth century and shifting relationships between sexuality, technology, commerce, and the law.

Information about these inflatable rubber dolls is admittedly hard to come by and requires some creative archival work. Unfortunately, any surviving specimens of the rubber women themselves would probably have dried out and crumbled decades ago, if any had been preserved in private collections. The dolls were also expensive and likely sold in limited supply, making actual examples even harder to come by. Therefore, this history of the femmes en caoutchouc has been pieced together instead from the textual artifacts that surrounded these dolls: advertisements for "intimate" apparatuses and rubber goods catalogs, as discussed in chapter 2, but also reports from the Paris World's Fairs, articles about the dolls' various run-ins with the law, and even songs written in their honor. Luckily, there is a surprising amount that we can learn by looking at the cultural traces left behind by the dolls—which appear to have been primarily manufactured and sold in France—through the quotidian Parisian print media published between roughly 1850 and 1920. What we find there is a story about the origins of sexual technologies, in their form as manufactured consumer goods, that is not about how men's desire supposedly drives technological innovation but rather about how technological innovation itself gives new form to sexual desire.

From the time when references to the femmes en caoutchouc first surfaced in the 1850s, to the early prototypes placed on display in the 1860s, to their wider sale by the 1880s, and eventually to changes in the rubber industry that shifted companies away from the manufacture of rubber dolls, the cultural life of the femmes en caoutchouc proves to have been a lively and storied one. Originally associated with a wondrous boom in everyday rubber products, the dolls later came to be seen as the stuff of contraband smugglers, corrupt politicians, and sketchy operators advertising illegal goods. Today, in the histories written about sex dolls and sex tech, the rubber women have been all but forgotten. Yet these dolls are crucial to understanding the actual origins and evolution of contemporary sexual technologies. They laid the groundwork for the later, more widespread sale of sex dolls, and they also played a key role in establishing a Western cultural imaginary around such dolls. The femmes en caoutchouc are not only the very first sex dolls (as much as anything can be). They are also the stuff of some of the very first sex doll erotica, the very first sex doll jokes, and the very first debates about sex doll ethics. They are the sexual technologies that set the stage for a lineage of sex tech to come.

Rubber as a Sexual Technology and the Impact of Obscenity Laws

The history of the femmes en caoutchouc is bound up in the history of rubber, both as a technology and as an industry. Far from being incidental, the fact that these dolls were made of rubber is critical to their place within larger histories of sexual technologies. Indeed, the case of the rise and eventual fall of the rubber women in France offers a valuable window onto larger technological shifts in material goods across the globe. As extraction and manufacturing processes changed, exemplified by the introduction of vulcanization, rubber came more and more into daily consumer life in Western Europe and the United States. This was exemplified by the growing proliferation of rubber goods items during the second half of the 1800s and the first half of the 1900s, a great number of which were designed for uses related to sex.

Much existing scholarship on the interrelated histories of rubber and sex has focused on the creation of contraceptive and prophylactic items by large companies operating in the United States, Great Britain, and Germany in the first half of the twentieth century—although, as the present work illustrates, the roots of rubber's relationship to sexual technologies also lie in an array of earlier, smaller rubber manufacturers in alternate locales.[1] Condoms are often central to these narratives. Writing in her book *Protective Practices*, a history of the prominent London Rubber Company founded in 1915, Jessica Borge explains the profound shift in condom production that was sparked by the invention of vulcanization.[2] Explains Borge, "The rubber condom industry began with the discovery of vulcanization, or the 'heat curing' of rubber in 1843–44 by Hancock (in Britain) and Goodyear (in America). This made rubber easier to manipulate and handle in the manufacturing process."[3] Vulcanized rubber was also put to many other purposes related to sex beyond condoms. In *Contraception: A Brief History*, Donna Drucker describes how the vulcanization of rubber, which "made rubber stronger, more heat resistant, and more elastic," also led to a rise in "dependable and commercially available contraceptive barrier methods for women," such as cervical caps (figure 4.1).[4] These shifts were themselves tied to larger historical factors. Borge writes that the first mass production of rubber contraceptives began in Britain in 1877, for example.[5] This corresponds with the Brazilian rubber boom, a stretch of time in the late 1800s when most of the rubber used for consumer and industrial goods was being extracted from the Amazon basin by European operators, often using the exploited labor of enslaved Indigenous peoples. This, as Carlin Wing has noted, explicitly ties the history of rubber—and by extension, I would add, the history of sexual technologies—to histories of colonial exploitation.[6]

Figure 4.1
An advertisement for cervical caps (listed as *pessaires*) that are "guaranteed unbreakable." Catalog for Société Excelsior, rubber goods and sex toy seller, 1905.

The use of shifting rubber technologies for the creation of items related to sex was also by no means limited to contraceptive devices. If we look to the catalogs of other early rubber goods companies that emerged in the mid-1800s, especially those in the business of "medical devices," we find that items related to sexuality were already central to this industry at the time of its very emergence. Consider, for example, an 1851 catalog for a Parisian rubber manufacturing company operating under the name Varnout et Galante, which lists addresses on both Rue du Faubourg in Montmarte and in Place Dauphine on Isle de la Cité, directly in the center of town (figure 4.2).[7] The catalog proclaims that it offers an array of apparatuses and instruments in *caoutchouc vulcanisé* (vulcanized rubber), all "made under the immediate direction" of one Doctor Gabriel.[8] Following a lengthy foreword explaining the exceptional qualities of this exciting new material, Varnout et Galante offers prospective customers nearly fifty illustrated pages of rubber devices. Although the catalog skirts overtly describing these items' sexual uses, many are clearly tied to sexuality or sex-related practices. Alongside numerous items for vaginal insertion and douching (which were likely used to help prevent pregnancy) and more than one device designed to compress or suction onto a breast (including an "artificial breast" to help with breastfeeding, a circular cushion that resembled a bicycle inner tube that could be inflated around a breast, and the ominously named "breast compressor"), we find rubber corsets with some distinctly fetishistic touches (figure 4.3).[9] In these pages, the line between medical devices and sexual devices, already a muddy divide in the history of sex toys, is intentionally further blurred.

Surviving documentation suggests that in the decades that followed, the number of Parisian sellers of rubber goods and related "curiosities" grew considerably. With this growth, such companies became more explicit about their sale of items designed for sexual purposes. In 1900, an illustrated catalog from the seller Maison L. Bador advertised "special apparatuses for men and women" made of "dilated rubber."[10] Here, the language of medical legitimacy (like the assurance that all rubber goods have been designed under the guidance of a doctor) has been replaced with the language of privacy. "With us," a letter to customers at the start of the catalog promises, "security and discretion are absolute."[11] Condoms are now prominently on display in this catalog, alongside vaginal "injecting and

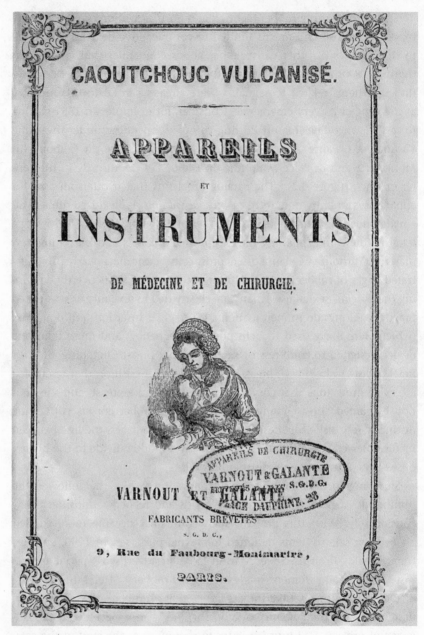

Figure 4.2
The cover of an 1851 catalog for "vulcanized rubber apparatuses and instruments," featuring a breastfeeding woman on the cover, produced by the Parisian company Varnout et Galante.

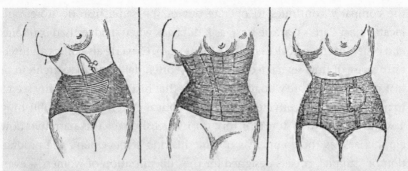

Les personnes qui n'habitent pas Paris, pourront, sans se déranger, avoir des bas et des ceintures faits sur mesure en nous renvoyant un des modèles ci-dessus remplis, ainsi qu'il est indiqué à la page 46.

Figure 4.3
A variety of rubber corsets for sale through a catalog of rubber medical goods published by Varnout et Galante in 1851.

dilating" devices so elaborate they resemble handheld blenders and a dildo with an extra attached nub for clitoral stimulation, the predecessor to the contemporary "rabbit."[12] This proliferation of catalogs also reflects a shift in the commercial landscape of sex toy sales in Paris—and therefore, because many of these catalogs were also mailed to nearby countries, a shift in the landscape of sex toy sales across Europe. Advertisements for these catalogs first begin appearing in Parisian newspapers in 1891, starting with the rubber goods manufacturer Maison A. Claverie, located at 234 Rue de Faubourg St-Martin.[13] Almost immediately after, many other companies describing themselves as *maisons* (literally "houses," a term typically used for those that produce rather than simply sell goods) start advertising alongside Claverie. This seems to prompt Claverie to start including warnings in their ads that other sellers are simply copycats offering inferior goods: "Do not trust crude imitations," these ads admonish customers, which are "sold by certain stores that only exist on paper."[14] Where exactly the devices sold by these other companies were being fabricated remains an open question.

Maison A. Claverie offers a particularly compelling piece of the Parisian rubber goods puzzle because, among the many companies that sold such items, they are the only one still in business today—more than 160 years after they were reportedly founded in 1860.[15] Now rebranded as Mademoiselle Claverie, a boutique that specializes in high-end corsets and lingerie, the company continues to operate out of the same historic storefront, located near Paris's Gare de L'est, as it did back when it advertised condoms and other intimate devices in 1891 (figure 4.4). Claverie also offers an interesting case of how sexual technologies are often deliberately written out of history, even by those who participated in that history. The company seems to maintain its own private archive of historical materials, even publishing a section about the store's history on its website. Yet this narrative now emphasizes, as the company's notable historic achievement, the production of "supple" corsets designed for the "liberalization of women"—even though earlier images of the Claverie factory clearly foregrounded other forms of production (figure 4.5). Thus, the way that Claverie tells its own story notably omits any references to rubber or sex-related items.[16] Even for a company that seems to have spent its first sixty years in the business of rubber, identifying as a rubber sex device manufacturer appears to be, today, undesirable.

In addition to the rise of vulcanized rubber and Parisian sellers of sexual devices, another major factor that shaped the history of the femmes

Figure 4.4
Illustration depicting the exterior of the Maison A. Claverie store, which is still in operation at the same address. Image appears in a 1919 catalog for "La Française," a style of prosthetic leg.

Figure 4.5
Illustration depicting the interior of the Maison A. Claverie factory. Image appears
on the cover of a 1900 catalog for orthopedic devices, which Claverie produced along
with sexual items.

en caoutchouc was French obscenity laws. The legality of the femmes en
caoutchouc and related sex toys remains murky, but numerous anecdotes
suggest that sale of the dolls was considered illegal, as was distributing
advertisements for such items. As discussed later in this chapter, there were
multiple court cases in the city of Paris involving charges related to the
sale of rubber women, as well as at least one work of erotic fiction that
describes the sale of sex dolls as illegal, though these records do not suggest
that consumers of such items were prosecuted for purchasing them.[17] Court
reporting on these cases notes that defendants were charged with *outrage
à la morale public et aux bonnes moeurs* (loosely, "offense to public morality

and correct behavior"). This charge would have fallen under a clause in the French penal code that forbade public indecency, which, broadly speaking, covered "any lewd act, solicitation, and exhibitionism," though the way this clause was interpreted shifted over time.[18] It was variously used to justify arrests for crimes such as engaging in sex with a person of the same gender or producing unauthorized nude photographs.[19] In her book *The Fallen Veil: A Literary and Cultural History of the Photographic Nude in Nineteenth-Century France*, Raisa Adah Rexer notes that police registers from the period document arrests as early as the 1850s for the sale of "illicit sexual objects," including "statuettes, paper dolls, pipes, snuffboxes, the odd *consolateur* (dildo), and condoms."[20]

The specifics of how and when this law against indecency arose in relation to sex dolls is hard to pin down. Censorship laws in France went through notable changes following Louis Napoleon's coup d'état in 1851, tightening considerably in 1852 and then loosening again in the early 1880s, as Rexer explains.[21] However, in conjunction with the femmes en caoutchouc, we still see the arrest of street vendors for the sale of pornographic materials as late as 1900.[22] This suggests that the dolls' legal status too, along with cultural notions of "morality" in relation to sex dolls, likely shifted over the period in question. Whatever the legal specifics at any given moment, this sense of the dolls as unlawful had an impact on the actual business of making and selling rubber women in Paris, making the dolls the stuff of euphemism and clandestine operations, while also forming a backdrop to many elements of the cultural life of the femmes en caoutchouc.

More broadly speaking, the legal tensions that surrounded rubber sex dolls in Paris also draw our attention to the broader impact of obscenity laws on the history of sexual technologies. Although they have varied across time and from country to country, such laws have played an ongoing role in both regulating and shaping these technologies. Consider the rabbit vibrator, which I mention again ahead in discussing contemporary rubber jelly sex toys. The rabbit has been now popularized in America, entering into cultural awareness through mainstream media appearances like the infamous 1998 vibrator episode of *Sex and the City*.[23] Popular histories of the rabbit written in the American context often focus on the supposed advent of animal-shaped vibrators in 1983, when the company Vibratex first began importing rabbit vibrators to the United States, where these "happier, friendlier alternatives to the standard vein-y penis dildo [appealed] to the ladies."[24] However, the rabbit's origins lie not in the United States but in Japan—where Japanese obscenity laws, which restrict the sale of items that too closely resemble genitalia, supposedly sparked the design of animal vibrators. Admittedly, as with the many origin stories for sexual technologies, the reality of the rabbit's origins is more complicated.[25] However, this example demonstrates how sexual technologies can shift in response to regulation, reminding us that it is not only moments of possibility and permissibility that have given sexual technologies their form, but also moments of restriction.

These notions of what is legally and culturally permissible are particularly resonant in the context of sex dolls. The fantasy that sex tech's origins might lie on foreign shores (whether in Europe or Asia) is a fantasy about what is permissible: a sense that exciting sexual adventures and inventions are possible elsewhere. Indeed, many contemporary texts that claim to survey sexual technologies and sexual cultures explicitly point to Asia, and more specifically Japan, as a place where anything goes—some imagined Orientalist mecca of sex tech where (men's) sexual desires are not encumbered by Western morals.[26] Sex dolls themselves play on worrisome fantasies of permissibility; they offer a vision of sex in which no permission is needed. History tells a different story. In the case of the rabbit vibrator, as for the femmes en caoutchouc and other sex-related items produced by the early rubber goods industry, we see creating items that appear to offer sexual freedom has actually long been a matter of navigating and adapting to forms of sexual control.

Les Femmes en Caoutchouc

Among the many items sold by Parisian rubber goods manufacturers in the second half of the nineteenth century and the first decades of the twentieth were the femmes en caoutchouc: inflatable, penetrable items made to resemble women's bodies. While other items from the period replicated just the genital area—and, in some cases, a *demi-corps de dames*, or a "woman's half body"—these dolls seem to have been relatively complete human forms.[27] Existing accounts suggest that the rubber women could be broken down into parts and folded up to fit in a box that looked like a suitcase.[28] Descriptions of these suitcases, wood- or leather-sided and unassuming, match similar traveling cases for vibrators and other sexual "medical" devices from the period. At least one surviving advertisement includes a note that the doll for sale has a head made of wax that could be customized according to a photo submitted by the buyer.[29] Accounts vary on this point, but it seems the dolls may have been able to be filled with a warm liquid, like water or oil. In some cases, it appears, this was done to help create a lifelike sensation for the person using the doll (making their rubber bodies resilient to the touch, like a waterbed); in other cases, the dolls seem to have been designed so that these substances could drip into their penetrable openings, as if they were vaginal fluids.

Reports suggest that the dolls were strikingly expensive, with prices mentioned ranging from 100 francs to 3,000 francs, the equivalent of roughly 300 to 9,000 US dollars today.[30] Some advertisements mention a variety of prices, depending on the size of the doll, a feature also found in advertisements for dildos, which could sometimes be purchased at a range of costs depending on whether the purchaser wanted the item in "small, medium, or large."[31] Although certain sources claim that the dolls were available in the forms of both women and men, most surviving materials only mention dolls shaped like women—but whether this means that dolls shaped like men were not manufactured, were less widely available, or were simply deemed too niche (or even unspeakable) to mention in surviving materials remains unclear.

Where and by whom these rubber dolls were actually produced is a similarly slippery question. It appears likely that, though there was a proliferation of rubber companies operating during the period, a much more limited subset manufactured and sold the dolls. As early as 1855, a French newspaper article mentions a rubber manufacturer setting up a new factory in Baltimore to make rubber women. Another such reference to a Baltimore rubber company that was supposedly manufacturing rubber sex dolls crops up again thirty years later, in 1885, in a novella in which a character makes passing reference to "the rubber women who are made in Baltimore . . . for amorous expansions."[32] This suggests that the femmes en caoutchouc, despite being more generally associated with French rubber companies, may in fact have first been made in America and then brought to Europe, perhaps via the major port cities of the US Atlantic coast.

A dive into the archival traces of mid-1800s rubber manufacture in Baltimore and the greater American mid-Atlantic and Northeastern areas brings the veracity of this suggestion into question, however. Records indicate that there were two established distributors of rubber goods in Baltimore in the mid-1850s. A Baltimore city directory published in 1851 includes a listing for Isaac Corbitt & Co., described as an "umbrella and India rubber goods dealer," and one for E. M. Punderson & Co., "manufacturers and dealers of umbrellas and India rubber goods."[33] Eight years later, it appears that Corbitt and Punderson remain the big players in the Baltimore rubber business. Two different Baltimore directories, each published as editions for 1858–1859, list "Isaac Corbitt" and "M. Punderson & Co." as the only notable local providers of rubber goods.[34] These records give no evidence that any new major rubber companies emerged in Baltimore around 1855 or that any Baltimore companies produced rubber sex dolls—though admittedly, the blanket term *rubber goods* leaves many possibilities open.[35]

Scouring these records does reveal other telling information, however. As we saw in the 1851 catalog from Varnout et Galante, items designed to interface with women's bodies and especially their breasts are frequently featured in the listings of early American rubber goods manufacturers and distributors. In the catalog for an 1850 Maryland exhibition of mechanical items, for example, we find a listing for a "gum elastic breast pump."[36] Around the same time, the US patent office records document two other patents for breast pumps, both of which emphasize their innovative use of rubber.[37] Thus, sex dolls were far from the only rubber good being developed during the period that were designed either to replicate or interface with women's bodies. Other examples, like a set of inflatable rubber breasts that could be strapped to one's chest using a piece of lingerie (advertised under the promise that they offered "an illusion of reality, not just in appearance but also in touch") further literalize the recurring presence of breasts in the early development of rubber goods (figure 4.6).[38] Ultimately, it appears unlikely that a Baltimore company truly did manufacture sex dolls as early as the 1850s, though it's notable nonetheless that French commentators were themselves drawn to this particular origin story for the femmes en caoutchouc.

Given the expense and size of the rubber dolls, as well as the issues of questionable legality surrounding their sale, I suspect that these rubber women were not among the more widely offered or better-selling sexual items to emerge from this boom in rubber goods. Yet this does not negate their historical and cultural importance. Like the sex dolls of today, the

Figure 4.6
Advertisement for inflatable rubber breasts worn separately or over one's own breasts. Catalog for Société Excelsior, rubber goods and sex toy seller, 1905.

femmes en caoutchouc were more than commercial products. They were also a cultural phenomenon, known to many people in the French context more through popular imaginaries than through contact with the dolls themselves. Consider, as a parallel, the place of sex dolls in society today. Contemporary sex dolls appear in innumerable works of twenty-first-century media, such as television and film, where they are represented in a range of ways: sometimes reverently, sometimes humorously, sometimes disparagingly, sometimes sympathetically. Even those among us who have never touched—let alone used—a sex doll are likely to have ideas about sex dolls. In much the same way, the femmes en caoutchouc lived a cultural life that extended beyond their actual use into the realms of everyday life and media, as we see ahead.

Rubber Women at the World's Fairs: Visions of the Rubber Craze in 1855

For roughly sixty years, from the 1850s to the 1920s, the femmes en caoutchouc appeared as recurring figures in news stories, works of fiction, and other materials printed in the robust network of quotidian publications that circulated through Paris.[39] Looking back across this period, we find that some of the most important historical nodes around which we see a broader cultural engagement with the rubber women are the Parisian *expositions universelles*: the five world's fairs that were held in Paris leading up to the turn of the twentieth century. These took place in 1855, 1867, 1878, 1889, and 1900 (figure 4.7).[40] Among their vast displays of goods, artworks, and indeed human beings (more on that in a moment), world's fairs famously foregrounded innovation, with displays of new forms of industry and technology presented as central attractions.[41] They drew millions of visitors to the cities across the world that played host to them.[42] As writing published around the Paris World's Fairs attest, such events were also hugely important events for the inhabitants of the cities where the fairs were hosted. With each world's fair lasting for roughly six months, they became a hub of local excitement, culture, and news during their runs. Throughout these times, as well as in the months leading up to and following the fairs, we see dozens of daily and weekly reports from *l'exposition* in Parisian newspapers and circulars. Thus, the world's fairs not only brought new technologies to their host cities but also placed these technologies

EXPOSITION UNIVERSELLE DE 1878.

PALAIS DU CHAMP-DE-MARS .

Figure 4.7
Engraved illustration of the Palais du Champs-de-Mars at the 1878 Paris World's Fair.
Published in 1878.

center stage in the cultural buzz of the day. Fittingly, this too is frequently
when the femmes en caoutchouc entered the French popular press.

If the world's fairs seem, at first glance, like unexpected sites for spotting
early sex dolls, it's helpful to remember that the fairs had a long and trou-
bled history of putting not only technologies but also human bodies on display. This includes sexualized bodies as well as racialized ones. Black bodies
and Black women's bodies in particular were presented as exhibitions at
many world's fairs, ranging from the 1893 Chicago World's Columbian
Exposition to the 1897 Tennessee Centennial and International Exposition
in Nashville to the First Portuguese Colonial Exhibition held in Oporto in
1934.[43] In these settings, the display of racialized bodies was often deeply
tied up with white, Western notions about the supposed exotic allure or
sexual availability of women of color, especially those from colonized
regions and the Global South—bringing together, as Isabel Morais writes,
"gender, sexuality, and empire."[44] Perhaps the most upsetting feature of
many world's fairs were the so-called human zoos, which turned the bodies

of women like Sara Baartman into curiosities deemed both abhorrent and alluring.[45] Gender and sexuality also played roles at the world's fairs in other ways, where shifting notions of women's labor were showcased in "women's buildings" and performances of masculinity were wrapped up with performances of technological prowess (figure 4.8).[46] In addition, the

Figure 4.8
The cover of an official illustrated guide to the 1889 Paris World's Fair, demonstrating the intersection of visions of masculinity and industry. Published in 1887.

world's fairs were quite literally centers for the sale of sex, whether in the form of sex work, sex toys, or pornography, precisely because of the massive crowds they drew from out of town.[47] Thus, far from being an unlikely locus in a history of sexual technologies, it makes sense that the world's fairs would serve as a backbone for appearances of the femmes en caoutchouc. It was through the fairs that technology, sexuality, and the body came most dramatically to the cultural fore.

The very first appearance of the femmes en caoutchouc in this archive of popular texts comes during the 1855 world's fair, which was held in Paris from May 15 to November 15. In September of that year, we find an article published in the newspaper *Le Figaro* as part of a regular column titled "Figaro à l'exposition" (*Figaro* at the fair).[48] As in each installment of the column, the article's author provides various vignettes and reviews of points of interest observed at the fair. After remarking on a surprisingly inexpensive new kind of furnace and praising an English artist's sculptural creations, the article takes on a more irreverent tone. One need no longer worry about accidentally smashing fragile objects, the author writes, because the Americans have begun replacing delicate materials like porcelain with rubber. Here, the author launches into an extended reflection—part tirade, part reverie, part comedic bit—about all of the rubber items that Americans have brought to display at the 1855 world's fair and many more that the author can only imagine:

> What am I saying, just porcelain? They have replaced lots of other things as well: objects for personal hygiene, small boxes, cabinetry, saddlery, sheet metal work, even cutlery. . . . What does this mean? It means that all human industry will involve this resinous gum, the veritable protean material of modern times. I suppose that colonel Colt's revolvers are now made of rubber; the gentlemen who pretend to dust showcases are made of rubber, with their rubber fans, on rubber glass. . . . That nation of rubber extends from Buffalo to New Orleans and soon to Cuba, from Baltimore to Texas and soon to Mexico. Rubber, what do you want from me? You frighten me but you reassure me. For the Americans, everything is rubber! Their navy . . . rubber! Their army . . . rubber!! Their nation's capital . . . rubber!!! Their democracy . . . rubber!!!! The measure of their soul . . . rubber!!!! . . .
> Rubber of rubbers, as the Americans say, everything is rubber![49]

In this frenzied description of an American rubber craze, we find hints about the actual array of rubber products displayed at the 1855 fair, as well as the cultural attitudes that characterized the reception of these products.

After listing items like silverware, saddles, and personal hygiene products among those that the Americans have purportedly constructed from rubber, the author takes the liberty of reimagining America itself as the land of rubber. Icons of America (at least in the eyes of Western Europeans) like guns, westward territory expansion, and even democracy itself are recast in a satirical insistence that, *chez les americains*, "everything is rubber!"

Yet the author of this *Le Figaro* article does not stop there in this exaggerated performance of disbelief. To illustrate just how far the production of rubber items has gone in America—or, rather, to illustrate just how far a French world's fairgoer might imagine the production of these items to go—the author turns to the subject of sex dolls. Addressing American readers, he writes:

> Utilitarian people, I admire you, but I also loathe you! . . . Will I no longer be able to throw my dishes out the window in order to hear the shards smashing on the pavement? Will I no longer be able to throw my wife out the window? Because now we are being assured that the Americans have decided to fabricate, for all of Europe, rubber women equipped with all the domestic virtues, offered at the fairest price, with all the graces of the enchanting sex, which we will be able to marry in three months' time with only the invoice from their manufacturer, and which we can throw in the boiler if we are not satisfied, satisfaction guaranteed![50]

Here, at the pinnacle of this author's list of absurd rubber creations, we find the femmes en caoutchouc. Whether the author is truly paraphrasing the promises of an American manufacturer (offering to provide the men of Europe with rubber women, "satisfaction guaranteed") is questionable. What we can say is that, based on this description, it appears that rubber sex dolls were not yet on display at the 1855 world's fair, though whispers about them had seemingly already begun to spread. This article also sheds light on newly forming public attitudes. Already, reactions to the rise of rubber were seen as inextricably tied to—and, indeed, epitomized by—sexuality. For better or worse, in the eyes of such commenters, rubber became both most bizarre and most intriguing when it stood in for women's bodies and when the properties of rubber (it can be bought and sold, it can be boiled down) sparked "jokes" about the assault of actual women.

Similar depictions of this imagined rubber craze continue to surface years after the 1855 world's fair, though they shift in their contents over time. For example, in 1880, the lyrics for a song titled "Le Caoutchouc" (Rubber) were

published in *La Chanson* (The song), a circular documenting new works by
Parisian songwriters.[51] The song begins with the following verses:

> Oh rubber! Rubber!
> Everywhere
> We order you, we buy you,
> Because, wherever we are,
> A rubber
> Tip
> Really can be used
> For anything.
>
> Good rubber,
> It merges, dissolves, flows.
> It can be woven, sewn.
> It stretches; to each his own!
> Whether in the form of a cheerful monkey,
> Or a scowling face, we mold it:
> Industry and art,
> Pleasure itself has its share.
> Oh rubber!

In these verses, we can see a lively, tongue-in-cheek celebration of rubber
not unlike the rendition found in the 1855 *Le Figaro* report. Now, however,
rubber is not the stuff of *them* (those strange Americans whose very souls
are surely made of rubber) but rather the stuff of *us* (the French who purportedly buy such products). In addition, rubber is now described less as
absurd and more as wonderful—lending itself, apparently, to all manner
of uses.

If the contents of this 1880 song seem less satirical than the soliloquy
found in the news story from twenty-five years earlier, it's likely due to two
factors. First, in the intervening years, a proliferation of actual commercial
rubber goods had come to be produced and offered for sale in Paris. Rubber
was now not just the stuff of world's fair exhibitions but the stuff of everyday life. Second, with the rise of actual rubber goods sales in France came
the rise of rubber goods advertising. In fact, immediately beneath the lyrics
for "Le Caoutchouc," we find that *Le Chanson* ran a sizeable and presumably paid advertisement for a rubber goods seller named Maison Larcher.[52]
The advertisement includes a list of rubber items for sale that sound strikingly similar to the items mentioned in the song itself, suggesting that this
song about rubber might be little more than a thinly veiled advertising

jingle. Maison Larcher even promotes itself as having won silver medals in the 1855 and 1867 *expositions universelles* (a common feature of such ads), reminding us yet again of the importance of the world's fairs in both the French industry and imaginary around rubber products, whether sexual or otherwise.

Yet it is the last lines of these opening verses of "Le Caoutchouc" to which I want to draw particular attention. The song lists the major arenas in which rubber is used: "Industry and art / Pleasure itself has its share." This suggestion that one of rubber's primary uses was for "pleasure" is taken up in one of the song's later verses:

> When traveling, rubber is
> The soft cushion, the hot water bottle;
> For the dashing coachmen
> It's the waterproof peacoat . . .
> A finger cot, a glove;
> It's the pink garter
> That evening, oh! happiness!
> Will delight a groom's best man.
> Oh rubber![53]

In these lines, with the image of the pink garter that brings delight, we see again how sexual items—and sexual desire more generally—are key features of both rubber manufacture and the cultural imaginary surrounding them. Indeed, this erotic insinuation is made even more explicit in a later, updated version of the same song that was published in 1898. This version, also titled simply "Le Caoutchouc," has been expanded to include three additional, notably bawdy verses listing other rubber creations. Among them are rubber panties, rubber babies, and rubber women (femmes en caoutchouc)—for which, the song's lyrics claim, "husbands would give thanks."[54] The evolution of this song also reflects a broader evolution in the association between rubber and sexuality, with the pleasures of rubber going from the stuff of speculation and humor to a very real share of the rubber industry and then to the center of an imagined world of sex and consumerism.

Sex Dolls on Display at the 1867 World's Fair

If rubber women were the subject of whispers and comedic speculation at the 1855 world's fair, they seem to have become a thing of reality by the time the world's fair returned to Paris in 1867. The fair, which ran from April 3 to November 1 of that year, once again whipped up a flurry of reporting and commentary in the popular press. This time, we find a report of an actual rubber woman on display at the fair. In a piece that ran in the newspaper *La Vie Parisienne* (Parisian life) on August 3, 1867, an author writes: "A rubber woman, which I had always imagined could only exist in utopia, can now be found at the fair; she can be taken apart and easily fits in a suitcase; what's more, as the flyer says, the color of her skin is indelible, because the entire mass of the material of which she is composed is imbued with it. The doll has extensive makeup, with eyes painted on that prevent these ladies from sleeping. I'd say it's a lovely invention, but really, what purpose could it possibly serve?"[55] This brief report attests to the presence of the femmes en caoutchouc in the flesh, so to speak, at the second Paris World's Fair—though it's unclear whether these dolls were officially on display as part of the fair's exhibits or sold in the environs of the fair. As is common to find in writing on the rubber women from this period, the author uses a mix of fantasy (I thought you could only find [rubber women] in utopia) and sly, tongue-in-cheek humor (but what purpose could they serve?). Other reports from 1867 confirm that a growing proliferation of rubber items aimed at consumers were on display that year. As one reporter wrote, customers purchasing such products at the fair would now be able to "brave the rain," a reference to rubber raincoats and umbrellas, "if not the ridicule" of being seen in uncouth rubber clothing.[56]

Although reporting from the 1867 world's fair would make it seem that rubber dolls were finally poised to enter the commercial market, the picture that emerges in the years between 1867 and the next Paris World's Fairs in 1878 and 1889 is not so clear. Later news stories do describe the sale of femmes en caoutchouc—including one such story published in 1885 in a French colonial newspaper in Algeria, suggesting that advertisements for the dolls reached as far as French-occupied North Africa.[57] Yet where the rubber women appear more prominently in this interstitial period is in a network of pop cultural, fictional, and artistic works. One example is the 1885 short story "La Femme du capitaine" (The captain's wife) that I discuss in chapter 3. Another is an 1868 cartoon in the *Journal Amusant* (Funny newspaper), which features an image of a woman in vaguely Greco-Roman robes using glass-blowing tools on two seemingly nude, recumbent men (figure 4.9). The caption reads, "*Jeune femme en caoutchouc soufflant des*

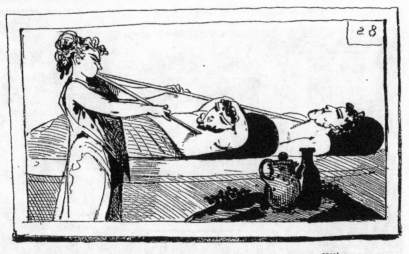

28. JEUNE FEMME EN CAOUTCHOUC SOUFFLANT DES MESSIEURS EN BAUDRUCHE,
par ALMA TADÉMA.

Figure 4.9
A satirical comic. This image appeared in an 1868 issue of the *Journal Amusant*.

messieurs en baudruche" (young woman of rubber blowing up men made of cecum), a reference to the bit of animal intestines commonly used to make condoms. The joke here seems to have two elements. First, the cartoon draws a humorous parallel between sex dolls and penises, both wrapped in rubber or a rubber-like substance. It also plays off the idea that the prospect of sex with an inflatable rubber doll inflated—so to speak—the aroused bodies of men. Examples like this of the dolls' appearances in media artifacts illustrate how, even as the actual manufacture and sale of rubber goods expanded rapidly, the true rise of the femmes en caoutchouc lay in their growing place in pop cultural depictions.

The cultural presence of the rubber women also manifests in more prominent work, such as George Méliès's 1901 short film *L'Homme à la tête en caoutchouc* (The man with the rubber head).[58] Méliès's film features a tinkering inventor who uses a bellows to inflate a rubber replica of his own head, delighting in causing it to expand and contract until his overeager assistant fills it with too much air, causing it to explode (figure 4.10). Although

Figure 4.10
George Méliès's 1901 *L'Homme à la tête en caoutchouc* (The man with the rubber head),
in which an inventor inflates a rubber version of his own head. *Source:* Screenshot
captured by author.

this film does not include overt references to sex dolls or sexuality, it
does feature a rubber body part that may well be a nod to the femmes en
caoutchouc of the day. Looking back, the dialog between Méliès's film and
issues of technology may already be clear, but the film takes on new, more
erotic meanings when we consider the other bodies en caoutchouc that
circulated through French media culture at the time the film was made.

Other work from the period more explicitly draws out the relationship
between fantasies of sexual technologies and the impact of the Paris World's
Fairs. One particularly striking short story, "La Femme en caoutchouc," ran
in the publication *La Tintamarre* in 1872.[59] It literalizes the *femme* in *femme
en caoutchouc* by telling the story of a man who accidentally marries a rubber doll. Apparently, he first notices that something is different about his
new wife when she bounces during sex. Eventually, she admits that she is
not a flesh-and-blood woman (she seemingly talks and moves around like a
human being), but rather a woman made of rubber bought from the 1867

world's fair and passed off as a person. "The only trouble," she tells her husband, "is that I am afflicted with certain inconveniences. You know how rubber expands with heat and shrinks with cold. . . . So don't be surprised to see me grow or shrink according to the temperature of the surrounding air."[60] Eventually, her husband grows to find his rubber wife irritating; when it gets hot in their apartment, he has to drag her out on the balcony and let her body compress in the cool air lest she take up too much space in the bed. To get rid of her, he poisons her, then melts her down and uses the rubber from her body to make a raincoat.

What is notable in this story, apart from its upsettingly casual misogyny (a recurring feature of so many stories about sex dolls across historical moments), is the way that it blurs the divide between rubber as a technology and the eroticized bodies of women. The character flaws of the rubber woman as a *woman* are one and the same as the material properties of rubber itself: because she expands and contracts at inopportune times, she inconveniences her husband, who in turn finds her no longer desirable as a romantic and sexual partner and so disposes of her. Above all, like the other examples from this period, the story demonstrates how the years surrounding the world's fairs in the 1870s and 1880s were ones when the purchase of rubber sex dolls was no longer seen as an absurd fantasy but rather as a reality. Surely this idea that a man could be duped into marrying a rubber woman is a flight of satirical fancy, yet the notion that such an item was available to purchase is presented by the author as known fact.

Brushes with the Law: The 1889 and 1900 World's Fairs and Beyond

Around the time of the next world's fair held in Paris, which took place in 1889, the cultural tides began to turn for the femmes en caoutchouc. Compared to articles from decades past, news reporting and other forms of commentary were comparatively quiet on the subject of sex dolls during the fair. There are a number of possible reasons for this. One is that the dolls may no longer have been seen as novel by the Parisian public. By 1889, a number of the rubber goods manufacturers mentioned earlier were already operating in the area, including ones advertising the sale of sexual items. Therefore, rubber dolls may no longer have held the same aura of curiosity and innovation that initially garnered them attention in the media.

More likely, though, this drop-off in press coverage can be attributed to the fact that the femmes en caoutchouc had crossed a threshold of public perception and were now associated with immorality and—as we see increasingly during this period—illegality. From the late 1880s onward, the rubber women become most visible to us when they come into conflict with the law, a common phenomenon in the archives of sexual history.[61] Like many artifacts of sexual cultures, their presence in official documentation is minimal. Yet once the law and other arms of governance get involved, a trail of documentation takes shape—particularly, for the rubber women, in courtroom records and news coverage of the day's courtroom dramas. It is here that we find the femmes en caoutchouc in the final years of the nineteenth century.

The first episode in the outlaw life of the femme en caoutchouc comes in the form of a museum of contraband. Three different articles, one from 1887, one from 1890, and one from 1897, all describe a "curious" cache of items, collected and stored at Paris's city hall, which police had purportedly confiscated from individuals involved in fraud, bootlegging, and the transportation of contraband goods.[62] Many of these items are unlikely and often entertaining containers that were reportedly used for storing secret supplies of items like alcohol, cigarettes, and lace, which bootleggers could then transport into the city of Paris and sell without being subject to local taxes. Some of these makeshift containers included corsets with hidden compartments, hollowed-out wooden logs, and fake babies whose stomachs were full of cognac.[63] All three sources make note of one item in particular, however: a full-sized rubber doll in the shape of a woman.

An 1897 article from the newspaper *La Lanterne* provides the most detailed description of the doll confiscated years earlier from *les contrabandiers*.[64] The author, writing about his visit to see the eclectic, dust-covered collection of items at Paris's Hôtel de Ville, provides a backstory for the doll and its use in the business of transporting contraband goods: "There [at city hall] we even saw a life-sized rubber woman. Every day she was dressed up like a bride, positioned in a very nice car, and her entire nuptials consisted of leaving Paris by one city gate and entering by another. The customs agent greeted the bride respectfully, and so she came into the city. She was filled with alcohol (*l'eau de vie*). The same system was used for a groom."[65] This description of the inflatable sex doll dressed like a bride and filled to the brim with illegal liquor is both ridiculous and surprisingly rich.

It links the rubber women not to the ingenuity of sex tech innovators, as we might expect, but rather to the cleverness of criminals. It transforms the second meaning of *femme* as wife into the literal woman in a wedding dress, who passes through customs checks precisely because the trappings of marriage make her so respectable that she is beyond reproach. Yet she is herself a contrabandier, one of the gang of fraudsters—an object, yes, but also an accomplice. And what about this groom? None of the other two reports about the museum of items at city hall mention a second doll. The groom doll could potentially confirm the existence of commercial dolls shaped like men as well as women, or it could have been a "female" doll dressed up in men's clothes. To think that two sex dolls, paraded around Paris like a newly married heterosexual couple, entered and exited the city limits sloshing with illegal alcohol . . . Such bold-faced instrumentalization of heterosexual privilege, rendered appropriately bizarre through the use of the rubber dolls, is almost as strange and charming as the tale of the dames de voyage itself.

By the time of the 1900 world's fair, the rubber women are experiencing their second major run-in with the law: a case involving the arrest of four hundred *camelots*, the hawkers who sold or handed out print materials on the streets of Paris. The camelots were central to the ecosystem of print culture in France and represented the primary methods of distributing these materials.[66] According to Jean-Yves Mollier in his historical study *Le Camelot et la rue*, thousands of merchants operating on foot sold goods in the environs of the 1900 world's fair, including many who sold sexual items like erotic postcards and—most importantly for our purposes—"rubber dolls strictly reserved for use by adults."[67] It was for the offense of selling these unauthorized erotic postcards, rather than the sale of sex dolls, that the hawkers were arrested en masse in October 1900, in the final weeks of the fair.[68]

Yet sex dolls proved to play an important part in this particular legal drama. Alongside the arrests of the camelots, Napoléon Hayard, a prominent if shady (not to mention nationalistic, chauvinistic, and anti-Semitic) figure referred to as "the emperor of the hawkers," was brought to court and accused of procuring the erotic postcards sold by the camelots.[69] Reportedly, Hayard insisted that he was not responsible for the influx of the postcards that had flooded the streets surrounding the fair. What's more, he said, he himself had been the victim of an entrapment plan. He claimed that minions of a local police commissioner had attempted to trick him into procuring an exorbitantly expensive inflatable doll referred to as a *femme ou une Eve en caoutchouc* (a "rubber woman or a rubber Eve").[70] Articles in at least two local newspapers, published on October 24, 1900, confirm that Hayard was indeed brought before the court on charges of moral indecency for selling lewd images.[71] Both articles highlight what one author describes as the "general hilarity" of the courtroom scene: apparently, when Hayard told the judge that the police had ordered him to get a rubber doll for "the bosses," the crowd burst out in laughter.[72] The reporter for *La Justice* describes the doll as a "bizarre accessory" and cannot help but add superfluous question marks to signal his amazement.[73] He writes, for instance, that Hayard "was asked by an agent of the prefecture to procure for him a rubber woman (??)."[74]

Hayard's brush with police (one of many over the course of his career, most related to the sale of obscene materials) highlights how the legal and cultural status of the femmes en caoutchouc shifted over time. Placed on display at the world's fair in 1867, twenty-three years earlier, their sale was now grounds for arrest. And yet in a different sense, the femmes en caoutchouc were still on display, whether as physical objects in the "curious museum" of contraband or as anecdotes in the courtroom. In these later moments, it would initially seem that public sentiment around the rubber women no longer envisioned them as a technology, but rather as a matter of legally questionable misadventures. Yet what the arrest of the four hundred camelots around the 1900 world's fair makes clear is that events like these were important nodes for the popularization of sexual technologies, whether or not those technologies were featured in the fairs themselves. The vendors who sold cheaply reprinted erotic postcards brought one form of sexual technology to the fair; the dolls that they reportedly also sold were another. In this way, the fairs can be understood as a site of concentrated erotics: a time and place when people had sex, purchased sexual items, and encountered sexual technologies. In this sense, the world's fairs, both in Paris and beyond, might be understood as their own form of sexual technology.

In addition, as the prominent place of the street hawkers in this 1900 run-in between rubber women and the law might suggest (as well as the prominence of advertising in the history of these dolls), the circulation of print materials played an important role in the cultural life of the femmes en caoutchouc. Two years after the incident around the 1900 world's fair, the rubber women were once again in the news, this time with reports of the arrest of a hawker named Monsieur Bouquet, who apparently ran aground of French obscenity laws when he distributed flyers advertising the sale of rubber women. In January of 1902, the newspapers *Gil Blas*, *Le Radical*, and *L'intransigeant* all published articles about the criminal charges brought against Bouquet, who was caught handing out flyers for "indecent objects"—among them femmes en caoutchouc, which were on offer for 2,000 francs.[75] Bouquet was charged with moral indecency and was sentenced to three months in prison. According to at least one of the newspaper accounts, the person who alerted the police about the flyer was a priest who had been handed a copy advertising the sale of a rubber woman.[76]

Bouquet's arrest for handing out flyers prompts us to reflect on the centrality of textuality and print media in the history of sexual technologies. In this case, the offense was not the actual sale of sex dolls but the distribution of advertisements. This is just one example of how the femmes en caoutchouc were as much a phenomenon of media as a phenomenon of actual material cultures. As for the dames de voyage, their history is almost entirely made of texts and tales. Information about these items, and above all information about how these items were perceived, comes down to us through articles, works of fiction, cartoons, songs: thus, as much as the technologies of rubber, it has been the technologies of print media that make this particular history of sex tech possible.

Rubber Today, Rubber Tomorrow

Following 1910, and especially with the start of World War I in 1914, the interest in rubber as a sexual technology began to fall away from French popular culture. A few examples of fiction featuring rubber sex dolls do appear in the years that follow.[77] However, these later works present the rubber sex doll as a thing of the past—items purchased "before the war" or increasingly outdated examples of sex toys from an earlier era.[78] During this time, Parisian rubber goods companies that had manufactured and advertised an array of sexual items, including rubber dolls, seem to have pivoted toward the production of goods related to the war and its aftermath, exemplified by Maison Claverie's shift from making condoms to making prosthetic legs for wounded soldiers.[79] And so, roughly seventy years after the figure of the rubber doll began appearing in the cultural imaginary of rubber and nineteenth-century sexual technologies, it comes to fade from view—leaving other items, like the plastic blow-up dolls of the mid-1900s, to take its place. Thus, despite their critical place as seemingly the earliest sex dolls manufactured and sold in Europe or the United States, the rubber women became largely forgotten items in the history of sexual technologies.

This does not mean, however, that the legacy of the femmes en caoutchouc ended with the decline of these dolls themselves. Indeed, the cultural and commercial life of rubber as a sexual technology persisted across the 1900s and does so up to the present day. The past few decades of the twenty-first century have shown us how the relationship between rubber and sex toys is continuing to grow and change. Along with it, the cultural significance of rubber is shifting—much as it did more than a hundred years ago around the Paris World's Fairs. The meaning of rubber, and its place within sexual cultures, is still very much up for debate.

Consider contemporary jelly rubber sex toys. Jelly rubber is a particularly floppy, translucent, and slightly sticky material commonly used for lower-end sex toys designed for genital penetration, including many versions of the iconic rabbit vibrator discussed earlier. Jelly rubber sex toys tend to be considerably less expensive than similar items made from silicone or glass. Over the last two decades, with the rise of feminist consumer discourse around sex toys exemplified by the corporate brands of companies like Good Vibrations, jelly rubber has become something of a stand-in for the dangers of cheap sex toys and the importance of buying quality items—part of what Lynn Comella describes in her study of feminist sex toy stores as "the politics of products."[80] Many commenters have described jelly rubber as hazardous because it can degrade inside the body, leaving behind harmful chemicals. A number of online articles bear titles like "How to Tell If Your Sex Toy Is Toxic" and "Yes, Jelly Sex Toys Can Be Dangerous."[81] By contrast, silicone toys, which are typically promoted as the safe alternative to jelly rubber, are often discussed by both consumers and marketers as more healthful and clean; here, silicone plays the role of the "good," new rubber in contrast to the "bad," outdated rubber of jelly toys. Now the question is no longer "What will they make out of rubber next?" (as many of the articles and ephemera of Parisian culture discussed in this chapter ask) but rather "What are the health ramifications, and indeed the feminist politics, of choosing between rubbers?" In both cases, the relationship between rubber and sex remains unsettled, changing, and up for question.

Rubber also holds an important place in the imagined sex tech of the future. The origin stories addressed throughout this book are typically deployed in the context of historical narratives that lead up to present-day "high-tech" devices, such as sex robots. Uncovering the history of the femmes en caoutchouc shifts our understanding of the factors and forces that have actually shaped the evolution of these technologies. Yet beyond that, these rubber women challenge us to attend to the material qualities of sex tech in the present day, not just in the past. Silicone rubber is key to the construction of increasingly "realistic" sex dolls by companies like Real Dolls, as well as many other items meant to replicate the human body either sold today or on display at highly visible events like the Adult Entertainment Expo.[82] It's rubber that gives these dolls their supposedly human-like look, feel, and movement. Thus, rubber itself, alongside its commercial history presented here, arguably represents the true origin point for contemporary sexual technologies—more so than any historical sex doll or sex toy.

Rubber itself holds a special place in our cultural imagination. Pliant and seemingly organic, its physical properties call to mind the sensations of the human body—like the floppy dick that Amanda Phillips links to alternative queer masculinities.[83] Rubber no longer strikes us as either the object of a baffling craze or a technological marvel, as it may have for readers and consumers in the late 1800s. However, even beyond its use for sex toys, it remains the stuff of alluring contradiction. Simultaneously unremarkable and intimate, banal and erotic, rubber is the material of our sex toys but also our everyday lives. Sometimes it projects out from us (as in the case of a strap-on) and sometimes it's inserted inside of us (as in the case of a dildo), but more often it is simply all around us in items we rarely stop to consider. The femmes en caoutchouc remind us that rubber itself, regardless of its use, is infused with a sexual history.

II Interrogating the Story of the Very First Sex Doll

5 Making Sex Tech Masculine, Making Sex Tech Straight: The Disavowal and Return of Femininity and Queerness

The origin stories we tell for sexual technologies set the terms for who and what matters at the intersection of sex and tech. This chapter represents a point of transition between the first and second parts of this book. Here, I shift from searching for the *dames de voyage* (or other, truer origins of the commercial sex doll) and begin turning a set of critical lenses back on the histories of technology that are already being told, exploring issues of gender, sexuality, colonialism, race, and the naturalization of sex tech. These are the concerns that have animated this book from the start; they bring research about the past to bear on the dominant forces of the present. Here in part II, I ask: What does it mean to tell the history of sexual technologies through a specific origin story like the tale of the dames de voyage? What cultural work does an origin story of this sort do, and what other possible stories does it obscure? Most intriguingly, what kinds of unintended implications do such stories bring along with them, casting the history of sex tech in a messier and even more radical light than those who tell these stories have intended? As I explained in the introduction to this book, origin stories of this sort merit both stringent analysis and subversive reinterpretation, part of a larger intellectual project of interrogating the ways that the intertwining histories of sexuality and technology are constructed and how those histories, both real and imaginary, can be reclaimed.

I begin with questions of gender and sexuality in this chapter because their relevance to the subject of sexual technologies is so immediate. In addition, this particular set of issues helpfully demonstrates the productive complexities of the cultural implications behind these origin stories. What the tale of the dames de voyage tries to communicate, in terms of gender and sexual identity, initially appears clear. As we have seen, discussions of sex dolls have long been bound up with sexist notions about women, their place within history, and the relationships that technology is supposed to have with their bodies. We have also seen how the politics of gender have profoundly shaped the path through which the tale of the dames de voyage has come into being, with women's writing being simultaneously central to and yet elided from the citational genealogies that underlie narratives about the history of sex tech. Although it has perhaps been less obvious, queerness has also been notably misrepresented or omitted by many narratives about the sexual devices of today. For instance, such narratives tend to celebrate the rise of technologies like teledildonics, sex toys that connect partners remotely via the internet or otherwise interface with computers.[1] Yet these same accounts conveniently neglect to mention that some of the key players in developing and popularizing teledildonics have been people like Kyle Machulis, whose work has often focused on the development of teledildonic butt plugs: sex toys that can be used by people of all genders but which are arguably more often associated with queer sexual practices than straight ones.[2]

Put simply, the tale of the dames de voyage, which I approach here both as a specific origin story and as a stand-in for the broader phenomenon of telling origin stories about sex tech, attempts to make the history of sexual technologies both masculine and heterosexual: the realm of straight men whose technological ingenuity can be attributed to their overwhelming desire to have sex with (something that at least resembles what they think of as) women. Yet these origin stories themselves—as well as their own origins, which lie in the trail of documentation I laid out in the first two chapters—often simultaneously resist this vision of history. This is exemplified by the dames de voyage in both of their imagined forms, as rudimentary sailors' dolls and elaborate proto sex robots. If we look just beneath the surface of their tale, we find that it actually upends a straight, masculinist vision of sex tech's origins, linking these origins to feminine-coded practices and imbuing the history of sexual technologies with a palpable homoeroticism. The same imagined story that attempts to make the origins of sex tech straight and masculine ends up making it femme and queer.

Making Sexual Technologies Masculine

In her 1999 book *Making Technology Masculine*, Ruth Oldenziel describes what she refers to as "men's love affair with technology"—a kind of "male technophilia" that manifests as a "passionate romance" that men conduct with computers.[3] Even when the passion that men have for machines comes under scrutiny from the wider public, writes Oldenziel, its logics remain unquestioned: as if straight, cisgender men's attraction to technology was itself "a matter of fact that needs no explanation."[4] Yet as Oldenziel explains, this notion of men's romance with machines has its own history. According to Oldenziel, that history starts in the final decades of the 1800s and runs up to the present day, emerging alongside shifting perceptions of gender roles and attempting to shore up a precarious, white, middle-class masculinity through narratives about the natural affinity between men and machines.[5] Mar Hicks, writing in their book *Programmed Inequality: How Britain Discarded Women Technologists and Lost Its Edge in Computing*, extends this explication of computer history, explaining how women were actively pushed out of the computer labor force (previously seen as women's work) in Britain in the 1950s and 1960s.[6] By the mid-twentieth century, women who worked in computational fields were being recast as low-skill workers, and less qualified men were being brought in to fill their jobs. This shift in cultural attitudes can be seen as its own kind of origin story, feeding into the erasure of women from other adjacent areas of computer history, such as the history of video game development, and ultimately seeding the male-dominated "toxic technocultures" that thrive online today.[7] In large part, these toxic technocultures are themselves founded on the presumption that digital spaces should be preserved as a realm for the unchecked expression of men's heterosexual desires.

All of which is to say that though technology has long been associated with men, that association is by no means "natural" or given. For at least the last 150 years, making it seem like technology is a masculine domain has required active cultural work. The feminist scholarship on computer history mentioned earlier is particularly valuable for making visible the *erotics* of coding technology as masculine. Part of the process of legitimizing men's claim to technological prowess has been the cultivation of a narrative about how men desire machines, both intellectually and sexually. This imagined "passionate romance," as Oldenziel terms it, has cropped up in popular culture for decades, since the introduction of technologies like the personal computer and the consumer internet. It appears in everything from how-to guides warning about the phenomenon of "computer widows," whose male romantic partners stay up all night rather than coming to bed, to Sherry Turkle's description in 1984 of young men who prefer computers to human girlfriends.[8] Scholars like Julie Wosk have demonstrated that the notion of men's attraction to machines, as it appears in art and fiction, also has a much longer history that continues to influence the present—emerging in works like E. T. A. Hoffman's "The Sandman" in 1816 and then in Léo Delibes's ballet *Coppélia* in 1870, which itself inspired Hans Bellmer's *Die Puppe* (The doll) series in the 1930s.[9] These are themselves some of the key works that have ignited dreams of sex with robots in the twenty-first century.

In its role as an imagined origin for contemporary sexual technologies, the tale of the dames de voyage actively contributes to this work of making technology and its history masculine. The story of the sailors' dolls, "invented" at sea by resourceful men hungry for the company of women, fits neatly into the vision of technology as the realm of men and their desires. Beyond this, though, the tale attempts to lay claim to the history of sexual technologies in particular (as opposed to technology more broadly) as an entirely masculine domain. By positioning the birth of sex tech in the supposedly all-male space of the sailing ship, the story makes it possible to tell the origins of sexual technologies as a story about men who desire women without any actual women in it.

As I mentioned in the introduction, there are many ways to tell the history of sexual technologies. Although masculinist narratives like ones that center the dames de voyage are increasingly common, these are matched by a number of important feminist histories, many of which focus on sex toys. Key examples include Rachel P. Maines's *The Technology of Orgasm*, Hallie Lieberman's *Buzz*, and Lynn Comella's *Vibrator Nation*, all of which root their histories in discussions of technologies designed for women's bodies, such as dildos and vibrators.[10] We could also add to this list works that understand sexual technologies to encompass technologies of contraception, such as writing by Donna Drucker or Claire L. Jones.[11] These narratives are neither simplistic nor overly idealistic about the place of women in the history of sexual technologies, recognizing that attempts to control women's sexual expression have often shaped sex tech history as much as attempts to empower them. Nonetheless, they can be understood as enacting a set of feminist politics around the telling of history that foregrounds women and their pleasure.

The tale of the dames de voyage and the approach to remaking history that it represents implicitly pushes back on these feminist narratives. While many of the books and articles that recount the story of the sailors' dolls mapped in chapter 1 do mention items like dildos and vibrators, they position the "very first sex doll" in a place of honor, often quite literally insisting that it (and not the "ancient" dildos that many popular histories of sex tech like to point to) should be understood as the origin point for the sexual technologies of today.[12] In this way, at the same time that such texts claim to celebrate the coming of a revolutionary sexual future, typically represented through the figure of the sex robot, the cultural work performed by the histories they tell is regressive. Even as they claim to present exciting new stories about sexual technologies, these accounts enact a reactionary renormativization of both technology and gender. This parallels Anne Balsamo's observation about more recent discourses around women as biotechnological cyborgs. "In many cases," Balsamo writes, "the primary effect of this technological engagement is the reproduction of a traditional logic of binary gender-identity which significantly limits the revisionary potential of new technologies."[13] Of course, the irony is that an origin story like the tale of the dames de voyage, which aims to push women to the margins, ends up recentering women in the form of the doll herself.

Indeed, if we look back across the citational lineages mapped in the opening chapters, we realize just how much work it has taken to shape the tale of the dames de voyage into a story about men. Presenting it as a tidy anecdote about straight male desire, far from being self-evident, has required effort to discipline the dames de voyage into their hegemonic role. The sexological works discussed in chapter 2, which form the historical "proof" for the existence of the dames de voyage in many contemporary texts, do not give the dolls pride of place or set them in opposition to sex toys made for women. Both Erich Wulffen in *The Sexual Criminal* (1910) and Leo Schidrowitz in *Supplement to the Moral History of Vice* (1927) describe the dames de voyage, which they understand broadly to be sex dolls or replicas of women's genitalia, as extensions of masturbatory devices for women.[14] Wulffen, for instance, mentions the dames de voyage as a side note in his larger discussion of "phallic fertilizers."[15] Sexual items for women also appear prominently in the Parisian rubber goods catalogs in which the actual rubber sex dolls of late nineteenth- and early twentieth-century France were advertised. Many of these catalogs in fact feature items for women (like contraceptive sponges and elaborate douching devices) more prominently than those for men. And a handful of now much-cited early texts, like René Schwaeblé's 1904 *The Wild Women of Paris* and Iwan Bloch's 1907 *Sexual Life of Our Times*, mention dolls shaped like men as well as women.[16]

Thus, writing the tale of the dames de voyage in the present day has required simultaneously legitimizing the tale by pointing back to older texts while also engaging with those texts in highly selective ways. Bringing the tale into being in its present form has necessitated overlooking elements of these texts that suggest that the dolls in question might in fact best be understood as technologies created alongside—or perhaps even used as—sexual devices for women as well as men (or devices used in non-heteronormative ways by people of various genders). This reflects attempts to remake the history of sexual technologies as masculine, but in addition, it also serves as a reminder of the fragile masculinity that in fact underlies the imagined "passionate romance" between men and machines. Apparently, women can be replaced with sex dolls, but it is unacceptable to imagine that men or their bodies might equally be rendered into objects used for the pleasure of others.

Looking back to these earlier documents also reveals just how much contemporary accounts have had to work to shift the characteristics (as well as the physical forms and the historical contexts) of the dames de voyage. Some of the most influential examples of these contemporary accounts specifically highlight the dolls' imagined inability to speak or protest, describing them as ideal partners—at least in the minds of the sailors who supposedly created them—because they remained eternally silent and compliant.[17] It's true that a number of primary texts, like a few of the short stories about the "rubber women" published in French quotidian newspapers in the late nineteenth and early twentieth centuries, similarly mention the purported appeal of a lover who has neither "whims" nor "migraines."[18] Yet there are also many striking examples in these precontemporary texts of sex dolls fighting back. We see this, for example, in the courtroom reporting on obscenity trials related to the sale of the femmes en caoutchouc, discussed in chapter 4, where men find themselves subject to legal punishments and public ridicule for their association with sex dolls.

In fictional works, the revenge of the sex doll takes a more literal form. Consider, for example, the 1899 pornographic novella *La Femme endormie*

(The sleeping woman) by pseudonymous author Madame B., frequently misinterpreted as proof of early mechanized sex dolls. Wealthy libertine Paul, the story's protagonist, has commissioned a local Parisian artist to construct an elaborate sex doll for his personal use. For much of the story, Paul engages in all manner of sex acts with the doll, following his own erotic caprices. Yet there are some moments when the doll herself seems to take on an alarming animacy. In one such instance, Paul approaches the doll with plans of embarking on yet another indulgent sexual escapade:

> All of a sudden, [Paul] stopped, stupefied. A cold shiver ran down his spine and he felt his hair rising.
> The doll had moved a leg.
> "Gads!" he exclaimed after a few seconds of panic, "am I batty or am I going batty? What is the meaning of this? She moved or I dreamed she did, or I saw double, or I've drunk too much—but no, I know that's not the case. Let's not be childish. Let's examine things carefully."
> He went to the side of the bed, leaned over the doll, and broke out laughing. "How ridiculous can one be? I was shaking like a four-year-old child."[19]

Moments like this interrupt the fantasy of the always-consenting doll, drawing out an underlying anxiety that the sexual object will animate and become a sexual subject. (Though paradoxically, this too is part of the fantasy of the sex doll, a figure frequently framed through references to the myth of Pygmalion, a sculptor whose beautiful creation comes to life.) A standard trope across stories about sex dolls, both historically and in the present day, is that they look, feel, or even move so much like human women that some character initially believes them to be "the real thing." Here, that realness bites back, revealing the masculinity associated with owning such a doll to be a fragile, frightened one. Faced with the possibility that the sex doll might take on autonomy, the lustful rake shakes "like a four-year-old child."

These examples illustrate how digging up the history of the tale of the dames de voyage does far more than allow us to confirm or debunk the story of the sailors' dolls. It also offers us the opportunity to identify how such a tale has been constructed both through and against earlier visions of the sex doll. The gender dynamics that characterize the long cultural life of the sex doll, it turns out, are far more complex than sex tech's origin stories, as told in the twenty-first century, would make them appear. Rather than bolstering the self-congratulatory notion that straight male desire sparked the invention of sex tech, these documents reveal how the "high-tech" masculine facade that surrounds both sexual technologies and technology more broadly began cracking long ago.

Laughing at Men: The Precarious Masculinity of the *Dames de Voyage*

Historical anecdotes about sex dolls may seem to be about women (women's bodies in the form of the doll, men's desire for women, the end of the need for women), but really they are about men. These anecdotes work to build a specific vision of masculinity, one that supposedly links the creation of sexual technologies with a robust, manly lustfulness exemplified by the image of the European sailor from days of yore. Although the story of the sailors' dolls is itself quirky, charming precisely in its eccentricity, its purpose remains serious. It's designed to convince those who might view contemporary sex tech in a skeptical light that these technologies have a long and venerable history and that they, along with the men who use such technologies or long to do so, should be taken *seriously*.

For contemporary texts that set out to tell the history of sex tech, getting readers to take sex dolls seriously is no easy task. US popular culture frequently presents sex dolls as humorous, deploying what Rebecca Clark has called a logic of "sex doll slapstick."[20] We might think here of the image of the iconic blow-up doll, made of thin, floppy plastic, lipstick-rimmed lips pursed into an absurd ring: a stand-in for the ridiculousness of sex dolls and a recurring prop in physical comedy mocking those who use them. Even when sex dolls and their users are portrayed more sympathetically, they are still, in one sense or another, "funny"—as in the case of the 2007 romantic comedy *Lars and the Real Girl*, which centers on a man in a romantic relationship with a lifelike doll. For Lars, personal growth means leaving the doll behind and "getting serious" with a flesh-and-blood love interest.[21] Countless popular films and television shows, from the various iterations of *Westworld* to 2014's *Ex Machina*, have approached questions of human-robot relations with a kind of hard-edged and often violent seriousness that is inextricable from masculine fantasies of technology, though some have interpreted this violence as an opportunity for antipatriarchal resistance.[22] (Granted, *Westworld* itself is not a monolith; the original 1973 film is far campier than viewers of the 2010s remake likely realize, with an aesthetic somewhere between that of a B movie and schlocky 1970s softcore porn.)

Yet the sex doll itself remains funny; it is only when she becomes so lifelike that her status as an object comes into question that she crosses over into the stuff of drama.

Central to the relationship between the tale of the dames de voyage and issues of gender is this effort to avoid mockery: the anxiety of a geek masculinity that is often seen, outside tech-focused subcultures, as in fact *less* manly and *less* virile precisely because it is so focused on machines.[23] The masculinities that have grown up alongside technology often prove to be defensive, retaliatory ones, enacting discrimination and even harassment against those who might, they fear, point a finger and make fun. As Katherine Cross has pointed out, vitriolic online campaigns like Gamergate can be explained in part by men's fear that women will "destroy" video games.[24] Yet we could equally say that this backlash results from a fear that men will be laughed at for the things they take pleasure in. This is relevant both in a realm of technology like video games that many still (erroneously) associate with childhood and also, and perhaps even more so, in the realm of sex.

Once again, however, if we return to the earlier sources on which the tale of the dames de voyage rests, we find that they undermine these very attempts to get sex tech taken seriously. Laughing at sex dolls and men who use them has in fact been part of the cultural conversation as long as sex dolls themselves, even before they were available as actual consumer products. It is telling, for instance, that a considerable percentage of the early published references to sex dolls from French quotidian publications appear in comedic newspapers. However, even texts published in more straitlaced venues pick up this thread of comedy. At times, such texts manifest their sex doll humor through an exaggerated performance of disbelief, like the author of the 1855 article who insisted that Americans had started manufacturing rubber dolls because, in America, literally everything was now made of rubber.[25] Yet in these documents, it's more typically the men who purchase or have sex with such dolls that are the butt of jokes, rather than the dolls themselves. These men are represented as dupes or oversexed buffoons, like the unwitting husband in the 1872 story "La Femme en caoutchouc" (The rubber woman) who gets hoodwinked into marrying a sex doll or the bawdy sea captain in the 1895 story "La Femme du capitaine" (The captain's wife) who cannot help but launch into boisterous, amorous entanglements with a rubber doll assembled from parts and transported in a suitcase.[26] Admittedly, in both stories, it may be men rather than women who have the last laugh. "La Femme en caoutchouc" ends with the doll's husband melting her down and making her into clothing, while "La Femme du capitaine" ends with the captain's actual wife embracing the sex doll to thank her for taking her husband as a lover.[27]

Other texts from this earlier period poke fun at men by casting them as cuckolds. For instance, an article that appeared in the August 27, 1885, issue of *L'Indépendent de Mascara* (a city in Algeria, one of France's colonial holdings) begins by describing how femmes en caoutchouc were now available for purchase.[28] Soon, however, it veers into a farcical anecdote about a doctor who asked the local train stationmaster to send away to Paris for a large male mannequin made out of rubber. When the item arrives, the Eve-like station master's wife becomes curious, opens the box, and finds herself standing before a "gentleman made of rubber, admirably well shaped."[29] She attempts to have sex with the mannequin, only to get stuck in its embrace. "Nine months later," the story concludes, "the station master's wife brought into the world . . . a rubber baby!"[30] Again, the joke is not on the sex doll—in this case a doll shaped like a man, rather than a woman— but rather on the people who engage with it: the doctor who orders it, whose pretext is flimsy at best ("he wanted to indulge in the opportunity to learn more about the elasticity of muscles"); the wife, seduced by a sex doll and now mother to a rubber baby; and above all the stationmaster, foolish enough to order the doll and to leave it where his wife, the Pandora of dashing rubber men, could find it. Now he finds himself father to a baby that is not his own and that, adding insult to injury, is made of *gomme élastique*.[31]

Curiously, many of these stories about sex dolls reflect a phenomenon of men laughing at other men. We might expect to find comedy used as a tool of protofeminist critique against dominant masculinist fantasies—an earlier iteration of what we see in contemporary American popular culture with examples like Whitney Cummings's 2019 stand-up comedy special *Can I Touch It?* "for the post-#metoo world," featuring her look-alike sex robot, or drag king Landon Cider's performance on the show *Dragula* as a sloppy, pizza-eating dude bro who has sex with a Real Doll.[32] In fact, in these early texts, it's far more common to find male authors poking fun at male subjects for the enjoyment of a presumed male readership.[33] This suggests that the origins of sex doll humor lie, at least in part, in the cultural policing that takes place between straight men: a way to insist on the "right" type of masculinity (the type that finds sex dolls absurd and does not "need" to buy a doll to have sex), as positioned in opposition to the "wrong" type of masculinity (the type that is so seen as desperate, so naïve, or so deviant as to have sex with a doll). Both of these types of masculinity are, in their own way, toxic.

The tale of the dames de voyage and the larger histories that it is a part of may be used to render today's sexual technologies serious—and, by extension, seriously masculine. Yet these histories themselves tell a different story about the kinds of masculinity that take shape around sexual technologies: a story about how the cultural imaginaries surrounding sexual technologies have always been bound up with humor, though the arrangements of who laughs and who is laughed at may differ. I do not advocate for laughing at other people's sexual practices, but I admit that there is a certain poetic justice in the fact that the very stories that are supposed to make us take the contemporary, male-dominated realm of sex tech seriously can actually be traced to a long textual history of laughing at men and their toys.

Making Sexual Technologies Straight . . . and Then Queer Again

As an origin story for sex tech, the tale of the dames de voyage doesn't only attempt to make the history of sexual technologies masculine. It also attempts to make this history straight. Time and again, in contemporary texts that recount the tale, we are told that the sailors who supposedly fabricated and used the first rudimentary sex dolls were driven by their lust for women—an urge that sprang from a deep-seated "germ of male desire," as Anthony Ferguson writes in *The Sex Doll: A History*.[34] The kind of straightness suggested by the tale of the dames de voyage is not just of any sort, however. It is characterized by a robust heterosexual drive—a hunger, even—for cisgender men to have sex with cisgender women, or at least their closest approximation. These stories also include homophobic elements, insisting that the idea of sex between sailors would have been so abhorrent that they were forced to fashion a woman-shaped doll.[35] Like the disavowal of feminist histories, deploying the tale of the dames de voyage is an attempt to disavow the possibility of telling a version of sex tech's history that centers queerness. Just as the masculinity of sex tech must not be funny, it must not be queer.

And yet, sex tech *is* queer. The actual place of queer and transgender people in shaping the history of sexual technologies is a rich subject that itself merits much more study. What I focus on here, however, is how the approaches to telling the history of sexual technologies that are being used to make sex tech seem straight actually end up rendering it palpably queer. Through the tale of the dames de voyage, we can see how efforts to disavow the queer implications of sexual technologies and their histories have in fact unintentionally highlighted the underlying queerness of these technologies. At the same time, positioning the story of the sailors' dolls as the origin of contemporary sex tech has actually made sex tech more queer, situating the birth of sexual technologies in a scene brimming with homoerotics and cultural visions of polymorphous perversity.

It is by now perhaps unsurprising to learn that when we return to the earlier texts related to sex dolls that later authors have used as evidence of men's supposedly timeless lust for women, we quickly realize that many of these works are anything but straight. For example, writing in 1903 in the first scholarly work to mention the dames de voyage, Iwan Bloch describes sex between men and dolls as an extension of homosexuality, claiming that the practice of making and using sexual dolls is closely tied to pederasty (sex between men and boys) and that most "statue lovers" are men who lust after other men.[36] Relatedly, Schwaeblé's 1904 Les Détraquées de Paris, which contains a short story about elaborate dolls made for sex and romance, is largely a text about women who have sex with women or who engage in butch displays of feminine masculinity.[37]

Once again, La Femme endormie provides a particularly striking case for consideration, this time of the close historical association between sex dolls and queerness. Wealthy protagonist Paul commissions his sex doll because, like Pygmalion before him, he has come to loathe the company of women—a sentiment with queer implications all its own. The doll he commissions is designed to look like a woman, but the story quickly takes a queerer turn. Not long after the doll arrives, the artist who made her breaks into Paul's home to have sex with his creation, kicking off a pornographic love triangle that persists for the rest of the story. Both men take turns visiting the doll, playing a coy game with one another in which each leaves his semen inside the doll for the other to find. On at least one occasion, Paul chooses not to clean out the doll before using her and instead revels in mixing together the bodily fluids of her two lovers. Soon, the men begin leaving intimate notes for each other, written flirtatiously as if from the doll's perspective. In one, Paul playfully chastises his sexual rival: "My darling, the last time you visited me, you forgot to wash me. You know that the state of my health prevents me from speaking or occupying myself with such menial tasks. So when you leave me after having had your fill of all the obscene things you love so much, please be so kind as to not neglect those duties of cleanliness that you yourself recommend. You'll find instructions for this operation on the stand. I kiss you, hoping to be kissed by you soon."[38] Fittingly, *La Femme endormie* culminates in a scene of group sex in which Paul and the artist take turns penetrating the doll. In these final moments, the two men engage in symbolic sex with one another, using the doll's body as a medium through which to connect and share an experience that is exciting to them, above all, because they get to do it together.

Yet nowhere are the homoerotic implications of the tale of the dames de voyage and its attendant cultural histories more prominent than in the story of the sailors' dolls itself. The various iterations of this story all place the origins of sex tech in a space that is clearly erotically charged: a sailing ship, packed with virile and sexually frustrated men, in some imagined European past. As I mentioned in discussing maritime studies and related archives in chapter 3, this vision of sailing ships from the 1600s to the 1900s as always entirely devoid of women is itself an oversimplification. Moreover, emphasizing the absence of women in the seafaring world serves to distract from the roles that women did play in maritime economies, such as in capacities as sex workers. My point then is not to say that Western ships from this period truly were all-male spaces. (After all, the tale of the dames de voyage is all over the map in terms of timeframes and nationalities. Which ships in particular would we even be talking about?) Rather, I want to highlight that contemporary accounts of the sailors' dolls expressly emphasize a vision of the seafaring world as populated entirely by men. A handful of these accounts even go out of their way to describe how these men surely slept close to one another in cramped quarters, lying awake at night as they fantasized about sexual partners and passed their homemade doll between them.[39] It's hard to imagine a scene thicker with sexual tension.

The queer implications of the tale of the dames de voyage lie in this vision of lustful men swapping sex toys, but also precisely in the figure of the sailor himself. Of all the archetypes of manliness to choose from, sailors are arguably the least closely aligned with heteronormative masculinity or straight male desire. Instead, across a wide range of pop cultural and artistic traditions, they have been figured as queer: omnisexual, hyperbutch and hypererotic, a staple of gay pornography and erotica. Sailors are the stuff of Jean Genet's infamous 1947 novel *Querelle de Brest* and Rainer Werner Fassbinder's 1982 film adaptation, *Querelle*, whose titular character is not only a smoldering bisexual sex worker, thief, and serial killer but also, first and foremost, a sailor (figure 5.1).[40] Sailors are also the stuff of Tom of Finland, of broad-shouldered, thick-necked men in leather embracing men in white sailors' caps with rippling muscles and visible bathing suit bulges (figure 5.2)—part of a larger midcentury queer culture of equating the aesthetics of the sailor, as Andrew Stephenson writes, with "popular mythologies about the sailor's erotic appeal and his voracious sexual appetite."[41] Sailors are the stuff of the Village People's 1979 song "In the Navy," with its spirit of "gay ecstasy."[42] And they are a recurring facet of queer urban histories, in which sailors both participate in actual same-sex practices and come to symbolize a kind of sexual freedom unmoored from geography, hegemonic social mores, and the constraints of heteronormative coupling.[43] The tale of the dames de voyage is an imagined history that hinges on a vision of sailors' sexuality. As such, it cannot be separated from a larger cultural imaginary that clearly figures sailors' sexuality as queer.

Figure 5.1
Movie still from the 1982 film *Querelle*, directed by Rainer Werner Fassbinder.

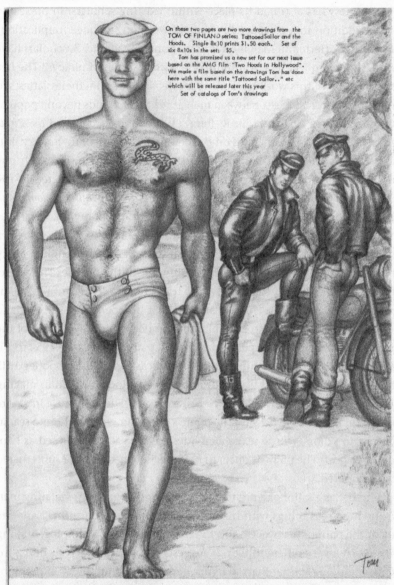

Figure 5.2
An advertisement for prints of Tom of Finland's drawings of the "tattooéd sailor" from a 1962 issue of *Physique Pictorial*.

Queerness has also had its place within existing scholarship on seafaring. In addition to a considerable body of work on the queer implications of writing by Hermann Melville, also mentioned in chapter 3, scholars have often explored queerness in conjunction with histories of piracy.[44] The historical nature of this work remains up for debate, but nonetheless attests to an ongoing interest in sex between men at sea that extends beyond popular culture and into academia. B. R. Burg's 1983 book *Sodomy and the Pirate Tradition*, for instance, operates on the belief that seventeenth-century British pirates in the Caribbean surely must have been having sex with one another, given (dubious) accounts that they seemed uninterested in the company of women.[45] Hans Turley, writing in 1999 in *Rum, Sodomy, and the Lash*, clarifies that actual "evidence for piratical sodomy is so sparse as to be almost nonexistent."[46] Yet he helpfully explains how the pirate as a cultural figure has long been bound up with queerness, even though many accounts of pirates' lives have left sexuality out of the picture. The pirate, writes Turley, is imagined as a "sexual transgressor" living in a "deviant homosocial world"; thus, "piracy and implicit homoerotic desire go hand in hand."[47] Turley points to examples that embody this queer vision of the pirate, like Cyril Ritchard's performance as Captain Hook in the 1954 musical rendition of *Peter Pan*, subsequently broadcast on NBC—a bootleg VHS copy of which, I admit, I watched on an endless loop as a child, making Ritchard's Hook the nautical icon closest to my own queer heart. Pirates are also useful figures here because the pirate's queer aura follows him to sea, in contrast to sailors, whose sexual activities are typically envisioned as happening in port. The pirate is queer in part precisely because he is untethered and unmoored.

Whether as a sailor or a pirate, the (sexual and sexualized) seafaring man on the high seas brings with him a distinctly queer valence that casts the tale of the dames de voyage—and the history of sexual technologies along with it—in a very different light. It suggests an origin for the sex doll born of queer longings rather than straight ones, offering a jumping-off point for reimagining sex tech's history as queer. At the same time, even read queerly, the tale of the dames de voyage sits in odd relation to queerness. It presents the origins of sexual technologies as both intimately entangled with queerness and yet also in tension with queer desires. Seen in this way, the tale suggests that what makes the sailors' dolls (as they are imagined) notable in the history of technology is not so much the inventiveness behind their creation as the dolls' own ability to ward off the looming threat of sexual contact between men. That is, acknowledging the queerness of the tale means reframing the drive behind the invention of contemporary sex tech as one characterized by homophobia and gay panic. Ultimately, the tale of the dames de voyage never fully resolves as either queer or straight. Instead, the form of sexuality it presents, like the tale itself, remains uncertain and caught in the balance: a heterosexuality perpetually on the brink of being subsumed into queer sex practices, a queerness always waiting to be jettisoned in favor of sex with a doll shaped like a woman.

Yet there are also other productive interpretations that a queer reading of the dames de voyage brings into focus. For instance, such a reading helps us envision the sailors' dolls themselves, as well as sex dolls as broader cultural figures, as a medium of erotic communication: a kind of uniquely embodied, predigital, preinternet teledildonics. As human stand-ins for penetrative sex passed among men, the dolls become a form of media—and a queer form of media at that. Namely, they become proxies, the medium through which men have asynchronous sex with other men. In this way, the figure of the sailors' dolls truly is a precursor to contemporary sexual technologies, though not in the ways that those who tell the tale of the dames de voyage typically imagine. It offers a vision of sex tech's origins that draw out the implicit queerness of men designing (and researching and promoting) devices for other men to have sex with—not in some distant past but in the sex tech landscape of today.

The tale of the dames de voyage itself, as we saw in earlier chapters, has been passed from hand to hand in much a similar way, moving not between lovers but between authors. In that sense, the tortuous citational lineage of the tale can be seen as a map of a kind of queer relationality among those who have written about the history of sexual technologies. As we have seen, this lineage is characterized by men citing or otherwise drawing from other men, sharing a vision of sex tech's origins among themselves while notably pushing the contributions of women to the side. Despite every effort of contemporary authors to paint a straight picture of the history of sexual technologies, the process by which those histories have come into being proves to have been just as (unintentionally yet undeniably) queer as the tale of the dames de voyage itself.

Femming the History of Technology

Much as attempts to make the history of sexual technologies straight backfire by rooting the origins of that history in queerness, the tale of the dames de voyage also undermines its own efforts to make the history of sex tech masculine. Seeing how the story of the sailors' dolls tries to disavow queerness while in reality rendering the origins of sex tech queer draws our attention to the fact that a similar contradictory dynamic is at play when it comes to gender. The tale of the dames de voyage, even as it attempts to render the history of sexual technologies masculine, ends up offering a vision of sex tech's origins that is surprisingly feminine—and, moreover, queer in its particular brand of femininity.

One of the ways in which the tale of the dames de voyage suggests reimagining the history of sexual technologies as feminine is by positioning the birth of contemporary sex tech at sea. In his article "The Genders of Waves," which explores shifting cultural associations between gender and the ocean, Stefan Helmreich writes that "seeing the sea as a feminine force and flux has a storied history. . . . The ocean has been motherly amnion, fluid matrix, seductive siren, and unruly tide, with these castings opposing such putatively heteromasculine principles as monogenetic procreative power, ordering rationality, self-securing independence, and dominion over the biophysical world."[48] In this sense, a form of technology that originates on the ocean carries with it cultural notions of femininity, tying it to mercurial moods, maternity, and the alluring call of the mythical creatures who inhabit the sea—who are, like the sailors' dolls themselves, both women and yet not quite human. As an origin story for sex tech, the tale of the dames de voyage offers a vision of the creators of "the very first sex doll" as men adrift far from the shores of solid ground. Here, solid ground functions as a metaphor for the prescribed "hardness" of male-dominated society, especially as it intersects with technology: the metals or machines of industry, the impenetrable black boxes of computation, and the world of masculine reason that supposedly underlies the "hard" sciences. The sea as a site for technology's origins offers an alternative with very different gender implications.[49] If we take this scene of birth seriously, then the sailors who invent the dames de voyage themselves become motherly figures, multifaceted in their gender roles, enveloped by the feminine-coded expanse of the open sea.

The tale of the dames de voyage also scrambles the history of sexual technologies as a history characterized by hardness—and, by extension, by normative teleologies of technological progress that are bound up with heteronormative masculinity. As noted, contemporary accounts of the dames de voyage oddly blur two ways of imagining these dolls: one in which the dame de voyage is a rudimentary sailors' doll made of scraps of cloth and leather and the other in which the dame de voyage is an elaborate, mechanized invention with complicated inner workings. Of course, as we now know, the actual early commercial sex dolls were neither of these things but rather "rubber women" whose bodies were pliant yet resilient: a combination, we might say, of both hard and soft. Together, these different manifestations of the dames de voyage, both imagined and real, offer a productively contradictory picture of the tangible qualities of sexual technologies and their gender implications. The vision of the sailors' dolls in particular renders technology soft. Made of fabric, stuffed with straw or cotton, designed to be embraced, she recalls a pillow or a homemade plush toy. Standing in contrast to the metal sheeting of machinery and computers, she has the potential to reorient the origins of sexual technologies (and technology more broadly) toward what Lara Mathis has termed *radical softness*. This suggests an opportunity to challenge the ways that the history of technology has been told by changing the way that it feels.[50]

Another way that the tale of the dames de voyage feminizes the history of sexual technologies is by presenting the creation of sex tech as a labor of crafting. The story of the sailors' dolls conjures up an image of manly men at sea patiently working with needle and thread to stitch themselves a sex doll. Tasks like sewing and weaving have, of course, long been considered "women's work"—though they too have played an important if often overlooked role in the history of computing, as highlighted through biographical writing on figures like Ada Lovelace.[51] And so, even as this particular origin story attempts to present sexual technologies as the domain of men, it conjures up other alternative narratives and counterhistories that center women and feminine-coded skills in the making of technology. There are also clear echoes between this image of sewing dolls as a form of technological innovation on the one hand and contemporary educational and critical making efforts on the other—efforts often led by feminist scholars and oriented in part toward encouraging more women and girls to participate in computing. Many of these efforts focus on e-textiles, wearable technologies, and sewing as an entry point into electrical engineering.[52] Even when it is deployed to serve a masculine history of sexual technologies, the tale of the dames de voyage suggests that feminine-coded work like crafting may have been at the heart of sexual technologies after all. The result is a vision of sex tech's first "inventors" that sits at a queerly fuzzy nexus of various gender identities and associations: hypermasculine men doing work associated with women, carefully fabricating a doll that will be both their creation and their lover.

The image of the lusty sailor diligently cobbling together cloth bits to construct a sexual companion portrays the act of technological innovation as simultaneously feminized and queer. Reconsidered in this way, the tale of the dames de voyage does not so much remake the history of sexual technologies into a women's history as it remakes it into a history that is enticingly femme. When I say *femme*, I am referring to what Andi Schwartz describes as "a queer identity marked by a critical engagement with femininity."[53] Others, such as Laura Brightwell and Allison Taylor and Rhea Ashley Hoskin, have argued for the importance of centering lesbian femme voices in gender studies and queer studies, calling for the development of a "femme theory."[54] In the present day, the femme—a kind of heightened, self-aware performance of femininity—is often tied to the digital, with social media platforms like Instagram being key sites of femme self-performance.[55] This suggests that contemporary digital technologies contain within themselves the potential to become tools for femme expression and that we could perhaps trace that potential back to earlier histories of technology that have led up to the present digital moment.

As an origin story with its own femme potential, the tale of the dames de voyage prompts us to ask: How might technology, both in its histories and its present forms, be *femmed*—reimagined as not only feminine but also queer in its engagement with and expression of femininity? In contemporary queer culture, we often speak of someone or something being *high femme*, a specific style of femininity that plays up this self-aware performance through the intentionally exaggerated use of makeup, hair styling, and clothing.[56] High femme both draws from dominant cultural notions of femininity and reveals their constructedness, much in the way that drag might simultaneously seek to embody "realness" while dismantling the notion of the real. We see a similar kind of performance at work in many of the accounts of the dames de voyage and other sex dolls discussed throughout this book, which often feature men carefully grooming their dolls, doll owners who have bought their dolls elaborate wardrobes, or dolls rendered lifelike through extensive makeup. The femininity of these dolls is, in its own way, a performance of the high femme. As for the men who are imagined to have made the sailors' dolls, they too participate in the making of a high-femme history of technology. They take their strikingly queer, strikingly feminine version of a supposedly straight male origin story and head out on the water—taking the high femme off to the high seas.

6 From Bamboo Lovers to Undersea Kingdoms: Colonialism and Race in Stories of Sailors' Sex Dolls

Like technology more broadly, sexual technologies are often imagined as separate from issues of race and the legacies of colonial inequity. Through the tale of the *dames de voyage*, we can see how colonialism and race in fact play a prominent role both in the contemporary imaginaries surrounding sex tech and in the histories that bring those imaginaries into the present. Sexual technologies, and sex dolls in particular, have always been inextricably bound up with colonial and racial violence, as well as fantasies about racialized and colonial subjects. Here, I shift from looking at how the story of the sailors' dolls attempts to make sex tech masculine and straight, the subject of the previous chapter, to interrogating the colonial and racial undercurrents that cross the story of the "very first sex doll." At times, through the dames de voyage and other concocted points of origin, sex dolls are imagined as themselves racialized: exoticized figures that embody white European and American fantasies about people of color. At other times, stories about early sex dolls specifically cast these dolls as bastions of whiteness. Colonialism too forms a palpable backdrop to the tale of the dames de voyage, which hinges on visions of European sailors embarking on seafaring journeys to colonial territories, off to do the work of empire. This chapter draws out the problematic colonialist and racial logics on which certain notions about sex tech's origins are founded; simultaneously, it brings attention to the ways that figures connected to the story of the sailors' dolls have been reimagined by the very people who have been the objects of colonial and racial oppression.

Far from being merely imaginary, the vision of the tale of the dames de voyage emerges from real histories of colonialism and racism—histories that strikingly parallel aspects of the tale itself—even as this story and others like it attempt to deny connections between these histories and contemporary sex tech. To illustrate this dynamic, I begin here not with the dames de voyage but with another tale: the story of the "Dutch wives." The story of the Dutch wives is an alternate origin story for sexual technologies, sometimes presented in contemporary historical accounts either in place of or alongside the tale of the dames de voyage. Prompted by the overt colonial and racial stakes of the Dutch wives, I turn back to the tale of the dames de voyage, identifying the racial and colonial elements of the tale as it exists in the present day, as well as in earlier texts that form its historical lineage. Drawing from the primary documentation discussed in previous chapters, I then demonstrate how cultural narratives surrounding sex dolls have long been structured around an implicit anti-Blackness. Building from this, I also discuss the echoes between the story of the sailors' dolls and real experiences of enslaved people during the Atlantic slave trade, looking to the work of Black scholars, fiction writers, and artists to make sense of the connections between histories of racial violence and the racialized imaginaries surrounding sexual technologies. I conclude by turning to related examples of Afrofuturist music, poetry, and visual art to think about how the tale of the dames de voyage might be radically reenvisioned in ways that build from rather than sidestep its ties to colonial and racial histories.

The Dutch Wives as an Alternative Origin Story

As I mentioned in the introduction to this book, the tale of the dames de voyage is just one of many origin stories—some more "real" than others—for sexual technologies in general and the sex doll in particular. These stories often intersect, and each helps draw out a different set of concerns underlying the construction of histories for contemporary sex tech. One such alternate origin comes to us as the story of the Dutch wives, which has much in common with and yet is also distinct from the tale of the dames de voyage. Like the dames de voyage, the Dutch wives have a long and complicated cultural history, which can be traced in bits and pieces across sources from the last few hundred years.[1] Similarly, they too raise questions about which elements of the histories of sexual technologies are true, which are imagined, and how imagined histories come to form. The Dutch wives merit their own extended study, which exceeds what I have space to present here. In lieu of that, I offer an overview of their story and its history, focusing on how this story brings to the fore the colonial and racial underpinnings that persist across various tales about sex tech's origins. For anyone interested in taking up the work of tracking down the Dutch wives, as I have done with the dames de voyage, I hope the information in the footnotes will offer a set of useful starting points.

The story of the Dutch wives appears in a network of contemporary American and European texts that overlaps with the textual network in which we find references to the dames de voyage. Although the specifics of these accounts vary, as we have by now likely come to expect, they largely portray the Dutch wives as early sex dolls originating—in one way or another—in East Asia or Southeast Asia and used by sailors in the Dutch East India Company, with relevant events taking place somewhere between the 1600s and 1700s (figure 6.1). According to some iterations of the story, the Dutch wives were sex dolls made by Asian people whom Dutch traders interfaced with in port cities (Japan's southern port of Deshima is a recurring locale in accounts of the Dutch wives, but Beijing, Indonesia, and Korea also make appearances).[2] There, men in the employ of the Dutch East India Company, a major colonial force that operated from 1602 to

Figure 6.1
An engraving depicting Dutch ships in Malacca Harbor in Malaysia. Image dates from 1676, the period in which sailors in the Dutch East India Company supposedly used Dutch wives.

1799, supposedly took a liking to the dolls and brought them back onto their ships, prompting the dolls to be later adopted across European fleets.[3] Other iterations of the story flip the account around, claiming that the Dutch wives were sex dolls originally made by Dutch sailors, who would "carry leather dolls around for the comfort of the crew"; the dolls were then adopted by the Japanese.[4] Still others claim that European colonialists like the Dutch invented the dolls after traveling around Asia and seeing woven bamboo cages used to help people in warm, muggy climates stay cool on hot nights, which gave them the idea for a related object that could be used for more sexual purposes.[5]

As with the tale of the dames de voyage, the story of the Dutch wives as early sex dolls is almost certainly a myth—a similarly muddled combination of historical fact and present-day fantasy.[6] The story of the Dutch wives emerges from its own winding, obscure citational lineage, but primarily it seems to be the result of the coming together of two semirelated elements.[7] First, in contemporary Japanese vernacular, the phrase *Dutch wife* (rendered as *dutch waifu*) is reportedly used for a variety of items, ranging from those of an explicitly sexual nature, like sex dolls and replica vaginas, to ones more open to interpretation, like large *dakimakura* body pillows.[8] Second, historically, the term *Dutch wife* has been used as an alternative term for the bamboo cages mentioned a moment ago, which are also referred to as *bamboo wives* or *bamboo brides* (*chikufujin* in Japanese, *jukbuin* in Korean, *trúc phu nhân* in Vietnamese, and *zhúfūrén* in Chinese; figure 6.2).[9] Why these items were referred to as *wives* is fairly clear, if suggestive: they are large, oblong objects designed to be brought to bed, where the user wraps their limbs around the cage so that cool air can flow between the open weave of the crisscrossed bamboo. As for how these items came to be known as Dutch wives, there are competing theories. Some scholars and commentators claim that it is because they were invented by the Dutch; some say it is because among European colonialists, the Dutch were particularly enthusiastic users of the dolls; and others claim that the adjective *Dutch* has actually been used historically as an insult, resulting from a period of colonial conflict in the seas surrounding Southeast Asia between colonial powers England and the Netherlands. This last possibility suggests that the *Dutch* in Dutch wives has little do with actual Dutch sailors and instead signifies that the dolls were "fake" or shoddy.[10]

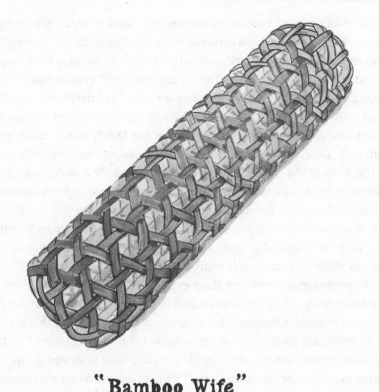

"Bamboo Wife"

Figure 6.2
Artist's rendering of a bamboo wife, a woven cage embraced like a body pillow to stay cool on hot nights. *Source:* Original art by Kathryn Brewster.

Whatever the reason, primary documents from as early as the 1880s written by white European colonialists living in occupied Asian territories do use the term *Dutch wife* to describe bamboo body pillows.[11] And though today the bamboo sleeping cages themselves seem to be largely considered relics in countries like Korea and the Philippines, they are still at times referred to as Dutch wives.[12] How Dutch wives came to refer to sex dolls and related items in the contemporary Japanese context remains an open question. However, if the evolution of the term *Dutch wife* is anything like the evolution of the term *dames de voyage*, I expect that it jumped from bamboo wives to sex dolls through advertising or other forms of euphemism. We have seen how, in the French context, sellers of sex toys used

similar rhetoric in attempts to evade obscenity laws. In Japan, where regulations against sex toys have persisted across the twentieth and twenty-first
centuries, it's reasonable to conjecture that similar factors may have shaped
the language of advertisements, introducing the use of a euphemistic, unassuming term like *Dutch wife* to signify a sex doll.[13] Whatever the truth, the
result is that more recent authors have conflated these various elements of
history and culture to produce the story of the Dutch wives, which envi-
sions early sex dolls used by European colonialists four centuries ago.

The story of the Dutch wives is itself quite literally a story about colonialism. It paints a picture of the cultural and economic exchange between
individual agents of European colonial superpowers like the Dutch East
India Company and local Asian communities as a *sexual* exchange. Whether
the Dutch are imagined as having taken early sex dolls from Asia or having
brought them to Asia, what is being passed between societies in this story
is the technologies of sex. Yet these exchanges, which position European
and Asian groups in different relationships to the act of invention, are not
imagined as equal. Contemporary accounts that state that the Dutch came
up with the idea for the dolls describe these sailors as "ingenious," couching them within rhetorics about white men and their imagined prowess in
realms of technological innovation.[14] By contrast, accounts that describe
Asian people as the creators of the dolls cast them not as innovators but
as representatives of long-standing cultural traditions: an implicitly orientalist, othering picture that paints Asian countries as home to titillating
ancient customs related far less to technology than to sex. Thus, in addition
to being about colonialism, the story of the Dutch wives serves as a window
into the sexualization of colonial and racialized subjects. It is also important that, unlike the tale of the dames de voyage, the story of the Dutch
wives takes place in ports, often associated in maritime scholarship with
prostitution, as I discuss in chapter 3.[15] The Dutch wife, as an imagined historical sex doll, can be seen as a stand-in for fantasies about white European
men's access to sex with Asian women.

Interestingly, this implicit link between Dutch wives and sexual fantasies
of colonialism is not just a feature of contemporary texts. It is one that we
see in writing from historical periods of colonial occupation as well. Long
before authors were claiming that Dutch wives were actually seventeenth-
century sex dolls, European colonialists played up the erotic implications
of the bamboo cages historically referred to as Dutch wives. In a letter dated

1882, later published as part of a collection titled *Travellers' Tales of Old Singapore*, an English subject writes home that bamboo Dutch wives are provided in all hotel rooms for European visitors.[16] "Don't be alarmed," the author cheekily assures the recipient of his letter, "it isn't what you are thinking of."[17] Dutch wives, he says, are simply objects to embrace while sleeping "for as much coolness as is possible"—though, the author jokes, drawing out the image of the Dutch wife as an actual lover, he does ultimately decide to kick the cumbersome "wife" out of bed.[18] An 1892 account from a French expatriate in the coastal region of what is now Myanmar, titled "A Sleepless Night in the Tropics," makes the erotic aura of the Dutch wife all the more explicit.[19] Describing the many obstacles to a good night's sleep faced by "European residents in Siam," he explains: "It was the month of April, when in temperate climes the 'young man's fancy lightly turns to thoughts of love;' but here our fancy is turned to ices, cold baths, and other cooling means, for April is the hottest month in the year."[20] Sweaty and frustrated, the author describes "rudely" tossing around his Dutch wife, but assures his "gentle reader" that it "is not a creature of flesh and blood, but only an inanimate bolster for resting the knees upon."[21] In both of these texts, the Dutch wife is portrayed as similar to an actual female companion who would appeal to men who lay awake at night, either from the heat or because their minds have turned to "thoughts of love."

Yet even as the Dutch wives represent the colonial and racialized fantasies that underlie visions of early sex dolls, their story also points toward the work of postcolonial subjects pushing back against these Western narratives. This becomes visible, for example, in Korean author Hwang Sun-wŏn's 1985 short story "My Tale of the Bamboo Wife."[22] In the story, a young man brings an older man a bamboo cage, curious as to what the item might be. The older man refers to it as a "naughty relic," telling the younger man that he does not need it because he is "still young enough to find the real thing."[23] Later he proclaims, "A bamboo wife keeps her purity no matter how old she is!"[24] Despite these innuendos, the bamboo wife in Sun-wŏn's story is not a sex doll or a sex toy. Instead, it is both something more banal (a bamboo cage) and more provocative (an invitation to imagination). Interweaving fables and magical thinking, Sun-wŏn takes up the figure of the bamboo wife and speculates poetically about the possibility that the bamboo wife could have been in the past—or might be in the future—a real woman. Sun-wŏn's story is itself deeply invested in the idea of origin stories, of which it proposes many, including one that posits that the very first bamboo wife was a figure in the shape of a woman made out of bamboo, presented as a concubine to the emperor of China. "My Tale of the Bamboo Wife" models how, when explored by a writer from a culture in which such bamboo cages were actually produced and used (e.g., Korea), the erotic overtones of the bamboo wife have the potential to take on meaning that is not contingent on histories of Western imperialism.

There remains much more to say about the Dutch wives, but I want to wrap up my discussion of these items by pointing to their reimagining in another work from Korea: the 2014 short horror film *Sleepless Night with Bamboo Wife*, directed and written by Seung-Ju Lee. In the film, a young man finds a traditional bamboo wife, with its iconic crossed bamboo slats, left out with the trash on the sidewalk and takes it home. His apartment is swelteringly hot and, at first, the item offers him much-needed relief (figure 6.3). Soon though, his relationship with the bamboo wife evolves into a sexual one when, in a dreamlike sequence, it seems to sprout arms and perform oral sex on him. Quickly the bamboo wife transitions from sensual lover to jealous monster. After the protagonist has sex with a female friend in his apartment, with the bamboo wife hanging in the background, the bamboo wife becomes semianimate, stabs his friend to death, seduces him, and then murders him. In the film's final scene, the protagonist's body is found, horribly contorted and bug-eyed, somehow enclosed in the bamboo cage.

Figure 6.3
Film still from *Sleepless Night with Bamboo Wife* (2014), directed by Seung-Ju Lee. The protagonist sleeps with the mysterious bamboo wife. *Source:* Screenshot captured by author.

Lee's film plays on the uncanniness, as well as the uncomfortable allure, of the bamboo wife. It is an object that is simultaneously intimate (it comes to bed with you) and dangerous—with its hard, diagonal edges and gaping holes, which are played up as appropriately menacing visual elements in the film (figure 6.4). Like the story "My Tale of the Bamboo Wife," *Sleepless Night with Bamboo Wife* demonstrates how the figure of the bamboo wife can become potent and complex when it is employed by Asian creators and decoupled from Western fantasies of sex tech's colonial and racialized past.

Colonialism and Race in the Tale of the *Dames de Voyage*

The prominence of colonialism and race in the story of the Dutch wives suggests the possible centrality of these issues in sex tech's other origin stories and prompts us to look for related issues in the tale of the dames de voyage. Indeed, both colonialism and race are highly relevant to the dames de voyage. Deployed as an origin story, the tale contributes to a larger project of colonizing and whitewashing the history of sexual technologies. By situating the origins of today's sex tech among European sailors, who may be from any number of countries depending on the account (France, England, Germany, Spain, the Netherlands, etc.), the tale suggests that what really matters is that the creators of "the very first sex doll" were not only men and not only straight, but also Western and white. This loose sense of the dolls' supposed pan-European origins displaces other possible

Figure 6.4
The woven pattern of the bamboo wife is an ominous visual trope in Seung-Ju Lee's film. *Source:* Screenshot captured by author.

places that we might situate the origins of sexual technologies, such as in non-Western or nonwhite cultures. Although "the East" and Japan in particular do figure into many contemporary accounts of sex tech's histories in the American and Western European contexts, as we have seen, these texts tend to represent Asia as an exotic land where the tech is "high" and the societal barriers to sex with machines are low.[25] In their discussions of Asian countries, such works tend to perpetuate orientalist visions of sexual and racial others, casting Asian technology as the object of white, Western desire rather than foregrounding Asian people themselves as key actors in the history of sex tech.

There are also direct ties between colonialism and the material histories behind the tale of the dames de voyage, such as in the production and circulation of rubber used to make the *femmes en caoutchouc*. In chapter 4, I explained how the rise of rubber goods (many of them sexual in nature) in the second half of the 1800s was tied to the Brazilian rubber boom and, by extension, to the exploitation of Indigenous peoples.[26] Yet the intertwined histories of colonialism and rubber sex toys goes back further than this, to the proliferation of India rubber, commonly used prior to the 1900s for a variety of sex toys, most notably dildos. Anjali Arondekar has described how the India rubber dildo, known for its particularly "realistic" feel, became an icon across pornographic and erotic pseudoeducational European texts—notable among them the Marquis de Sade's 1795 *La Philosophie dans le boudoir* (translated as either *Philosophy in the Boudoir* or *Philosophy in the Bedroom*).[27] India rubber, as Arondekar explains, was a product of colonial India—but it was part of a set of ideas about India that circulated through the European cultural imagination. In England especially, the India rubber dildo came to represent a conflation of Victorian sexuality and the reach of empire. At the same time, notions of the India rubber dildo shifted the sexual expectations placed on colonial subjects. Through the India rubber dildo, writes Arondekar, "Indian male sexuality in the nineteenth century is both emptied and reimagined."[28] That is, in the European imagination, the figure of the India rubber dildo set the terms for Indian male sexuality itself. Like the story of the Dutch wives, the story of the India rubber dildo highlights the role of colonialism in the making of Western sexual imaginaries.

Fittingly, if we return to the primary documents that reflect earlier histories related to the dames de voyage, we find that colonialism has long been part of the imaginaries surrounding sex dolls. This manifests, for instance, in the 1921 short story "La Poupée" (The doll), discussed in chapter 3, which featured a French traveler who had recently returned from Africa and was now showing off his lifelike rubber sex doll to an incredulous friend back in Paris.[29] In this story, the fantasy of empire is bound up with the fantasy of the sex doll herself, who becomes a handy companion (and an accessory) for colonial travel. The colonial undercurrent in the history of sex dolls also manifests, decades earlier, in a 1885 article in *L'Indépendent de Mascara* titled "Les Femmes en caoutchouc" (Rubber women), which announces to French expatriate readers that they can now send away to France to purchase a sex doll "with all the physical and moral faculties that a faithful and 'natural' wife brings with her into marriage." "I don't know if these dolls can cook," says the article's author, "but I'm sure they can do other things."[30] With this example, we see how the historical figure of the sex doll was not only shaped by colonialist fantasies but was herself a product that could reportedly be shipped to European colonies, bringing a bit of France with her to the French in Algeria. US settler colonialism comes into play too in these documents, such as in the 1855 article in *Le Figaro* reporting from the first Paris World's Fair, which insisted that Americans were so obsessed with rubber that they were now making rubber women—and next would come rubber Colt revolvers with which to conquer the Western territories.[31] Thus, colonialism is more than a backdrop to these texts: it is central to the fantasy, rooted in history yet persisting into the present day, that ties sex dolls to distant voyages and far-off lands.

The story of the Dutch wives also prompts us to reconsider the role that race plays in the tale of the dames de voyage and its related histories. Contemporary narratives about sex tech's origins tend to sidestep issues of race, as if the desire for sex with machines were itself an expression of a techno-utopian postracialism. If you are having sex with dolls (or robots) instead of people, these narratives suggest, racial politics no longer matter. Of course, twenty-first-century sexual technologies and the logics that surround them are in fact deeply tied to race. As scholars like Louis Chude-Sokei and Catherine A. Stewart have demonstrated, there is a long tradition of intermingling cultural imaginaries around Blackness and robotics in America, such as in connections between blackface minstrelsy and imagery of automatons or attempts to construct a robotic workforce of "mechanical negroes."[32] Similarly, when we look once again to the earlier lineage of texts that underlies the tale of the dames de voyage, we find that race has actually been a recurring concern in the cultural conversations and commercial industries surrounding sex dolls. The very same reference points that are being used to tell a supposedly postracial history of sexual technologies in fact document the centrality of race in these histories.

In particular, these documents attest to a long-standing thread of anti-Blackness, whether explicit or implicit, that pits the whiteness of sexual technologies against the imagined threat of Black sexuality. Once again, the 1921 story "La Poupée" serves as a potent illustration. When the French colonial explorer explains to his friend why he purchased the doll, whose "luminous" and "rosy" white skin is emphasized throughout, he states bluntly, "I detest Black women"—implying that his rubber companion would save him from having sex with nonwhite women while on his extended trip to Africa.[33] As if the racism of this statement were not blatant enough, the story actually ran in the newspaper La Vie Parisienne accompanied by a series of illustrated images of a Black man peering through an open window at the naked sex doll, who reclines seductively in the manner of Manet's Olympia. Presented as lecherous and primitive, his face pressed up against the window like a hungry animal, the man leers at the doll's expanse of white skin. In this way, "La Poupée" reflects how the racist attitudes that shape these fantasies of the sex doll implicate both Black women—through a sexualized version of misogynoir, a phenomenon addressed by scholars like Moya Bailey—and Black men, presenting the former as so unattractive that sex dolls must be manufactured to take their place and the latter as insatiably drawn to the imagined beauty of manmade whiteness.[34]

These ties between anti-Black racism and constructs of masculinity also surface in the material histories of the sex doll. Many of the advertisements for rubber goods discussed in chapters 2 and 4 go out of their way to emphasize the color of the "intimate" items they sell. This is especially common in advertisements for condoms. For example, the catalog dated 1900 from Parisian rubber manufacturer Maison L. Bador opens with a full-page ad for préservatifs pour hommes, en caoutchouc dilaté blanc ou rose—rubber condoms for men, offered in either white or pink (figure 6.5).[35] A few pages later, we find another prominently displayed advertisement for men's condoms made out of baudruche blanche incassable—supposedly unbreakable condoms constructed from animal intestines (which were often used for condoms before the rise of vulcanized rubber and later latex) that are,

Figure 6.5
An advertisement for condoms in white or pink rubber from a 1900 catalog of rubber goods produced by the Parisian seller Maison L. Bador.

the reader is assured, white.[36] These promises fend off other potential visual associations with such materials. Many (though certainly not all) vulcanized rubber products during this period were brown or black, as we saw in the image of the inflatable *ventre de femme* presented in chapter 2. Sheep intestines similarly had to be extensively processed to transform them into condoms that were not only strong but also light in color.[37] Thus, the rhetoric of these advertisements can be understood as an attempt to assure potential customers that the sexual devices they purchase, especially those that are placed like a second skin over one's genitals, will be neither black nor brown.

Simultaneously, advertisements for sex-related products from this period reveal additional layers of complexities and contradictions surrounding white European attitudes toward Blackness and sexuality. At the same time that rubber goods manufacturers promised customers that using their condoms would not make their genitals appear dark, other operators in the same commercial ecosystem were playing up stereotypes of Black hypersexuality to sell their wares. For example, in an 1895 issue of the circular *La Grisette*, an ad for condoms produced by the Parisian rubber goods manufacturer Maison A. Claverie runs just before an ad for "Nubian pills" and another for "Nubian injections."[38] These items, sold by companies with names like Pharmacie Speciale (the special pharmacy), insist that such pills and injections will make those who use them more "energetic." They will also, according to these ads, "stir up the blood and remove impurities" caused by sexually transmitted diseases like syphilis.[39] What makes these items "Nubian" the so-called special pharmacy does not feel the need to explain. Here, we see how Blackness (which is often conflated in these texts with African-ness) was anxiously avoided in some realms of sexual consumer goods and problematically appropriated in others. In these ads, "Nubian" stands in for a shared cultural understanding among the clientele of urban Paris that Blackness could be equated with promiscuity and sexually transmitted diseases but also a desirable sexual potency. Clearly, for more than a century and surely longer, the development and sale of sexual technologies, whether for pleasure or for quackery, has never stood apart from race. Rather, these technologies, both in their material forms and in their cultural reception, have always been characterized by a tense mix of racial panic and racial fetishization.

A Doll for Captain Pamphile: Echoes of the Slave Trade

Issues of race recur in the earlier documentation that leads up to the contemporary tale of the dames de voyage—yet the tale itself, as it is repeated in the present day, itself already contains striking echoes of histories of racialized violence. In particular, the image of the dolls transported on long sea voyages, during which they are used for sailors' pleasure and stored in the holds of ships (as multiple accounts posit), cannot help but call to mind the actual experiences of enslaved people during the Atlantic slave trade and the Middle Passage.[40] In her book *Slavery at Sea*, Sowande' M. Mustakeem explains the historical context that gave rise to the Atlantic slave trade: "the eighteenth century bore witness to a dramatic transformation in commercial slavery across the Atlantic that created a spiraling intensification for African laborers," resulting in an explosion in "the shipment of men, women, and children . . . deposited into various Atlantic ports and slave societies."[41] Forcibly transported to areas like America and the Caribbean, these enslaved people endured monstrously inhuman conditions, including being chained in impossibly small and unsanitary spaces in the holds of slaving ships during the journey across the Atlantic Ocean— the Middle Passage.

The sexual abuse of Black women formed a constant backdrop to the operations of the Atlantic slave trade. Mustakeem describes how both aboard ships during the journey across the Atlantic itself and while in port, Black women were often "loaned to . . . white sailors for sexual purposes."[42] Sexual assault of women captives by crewmen was itself a pervasive feature of life during the Middle Passage, as Mustakeem explains: "Regardless of age, females were regularly exposed to violence as seafarers sought to hold complete dominance over their personal lives. Deemed docile and thereby voiceless, black women had few if any methods to escape sexual violation. Instead, crewmen were 'permitted to indulge their passions among them at pleasure' throughout the passage. . . . With brute force a staple feature of slavery at sea, bondpeople, especially females, traveled the Atlantic, as . . . 'sexual hostages' often to more than one crew member."[43] This passage hauntingly parallels contemporary accounts of the sailors' sex dolls. In Mustakeem's description, we can see elements from these very real histories that also appear in writing about sex dolls across historical moments—such as the perception of Black women as "voiceless and docile" (traits that are often idealized in descriptions of sex dolls). The status of women captives during the Middle Passage as "sexual hostages" abused by multiple crew members also clearly recalls descriptions of the dames de voyage as objects shared among sailors as a way to indulge their own passions.

In drawing this parallel between the tale of the dames de voyage and the experiences of enslaved people during the Middle Passage, I am not trying to suggest that we understand the actual women who suffered these abuses as somehow similar to sex dolls. Such a suggestion would disrespectfully diminish the immense seriousness of the human suffering represented by the Atlantic slave trade. Instead, I am arguing that the story of the sailors' sex dolls, as it is repeated in contemporary accounts, is itself structured around (while simultaneously trying to disavow its connection to) interpolated visions of this history. Reconsidered through this juxtaposition, the oft-repeated anecdote about how the first sex dolls were made, stored, and shared by white, European sailors reveals itself to be far less "quirky" and far more worrisome. It changes the story from a tale about straight men's technological inventiveness and reveals it to be, at least in part, a thinly veiled fantasy about controlling the bodies of women of color—repurposing and contorting imagined scenes from the enslavement of African people, as well as the capture and abuse of others, such as the Korean "comfort women" forced into sexual slavery by the Japanese Imperial Army in the period surrounding World War II.[44] Seen in this light, the tale of the dames de voyage emerges as a tool in a larger white supremacist project of legitimizing not just the use of sexual technologies but also, by proxy, the abuse of real racialized subjects treated like sexual objects. In doing so, it dangerously transforms a scene of suffering into a scene of bawdy pleasure, removing concerns about consent and attempting to distance itself from its racist implication by reimagining captive women as literal dolls.

Yet this connection between the cultural imaginaries of the sex doll and visions of the Atlantic slave trade itself has a long history. If we look once again to the lineage of works that brings us the contemporary tale of the dames de voyage, we find that it has been tied to fantastical notions of slavery for at least the last 120 years. The key to unlocking the history of this connection lies in the 1899 erotic novella *La Femme endormie*.[45] This text, as I explain in chapter 2, is one of the foundational works in the genealogy of the dames de voyage, cited by numerous authors across the 1900s and up to the present day.[46] I have explained various elements of the novella's plot elsewhere in this book. However, it is of particular importance that before launching into the story of Paul and his elaborate mechanical lover, the author introduces the idea of the sex doll in this way: "A kindly soul has invented the dildo for women deprived of male contact; for the pleasures of brave Captain Pamphile, someone had brought forth the rubber woman [*la femme en caoutchouc*] . . . For our hero, a deft craftsman, an artist, would invent a miraculous Phrynée he would be able to manipulate as well—she would always be compliant and silent, no matter how lewd the act he chose to perform."[47] There are multiple elements we could point to in this brief passage that relate to arguments made across other chapters: such as how the author begins this list of sex objects from across history with those designed for women rather than men, or how the doll is described as a "miraculous Phrynée," a quick, unexplained reference to the story of Phryne, a Greek courtesan from the fourth century BCE, one of many instances in the citational history of the dames de voyage when sex workers are rendered both present and absent. However, the detail that I want us to pay attention to is the assertion that rubber sex dolls have been created "for the pleasures of brave Captain Pamphile." Referenced within this story, a text that is often cited as proof of sex tech's origins, we find another origin story—the suggestion that sex dolls were invented for a particular sea captain.

Captain Pamphile is not a real historical figure, however, but rather a fictional one—the eponymous character in the 1839 novel *Captain Pamphile* by Alexandre Dumas.[48] Nor do sex dolls appear in Dumas's novel, a reader of the book reveals, which itself predates the earlier documents discussed in this project by the better part of two decades. Instead, this reference to Pamphile in *La Femme endormie* should be understood as a stand-in for the figure of the sea captain more generally—and, by extension, the broader archetype of seafaring men. Yet if we turn to consider Dumas's novel in more depth, we find that Pamphile is not just any kind of sea captain. He is a liar, a trickster, power-hungry, ridiculous, always off on some horrid adventure: like a vicious, hot-blooded Baron Munchausen. He is also an unabashed racist, an agent of white colonialism, and a literal slave trader.

Indeed, a closer consideration of *Captain Pamphile* reveals just how deep this imagined tie between sex dolls and the escapades of "brave" European men who enact racialized violence runs. Dumas's novel is an unusual one, made up of equal parts absurdist travel diary and fin de siècle decadence. It simultaneously has overtones of curious predecessors like *Robinson Crusoe* (1719) and the nihilistic, casual cruelty of Sade's libertines; it also sets the stage for later works as seemingly disparate as *Moby-Dick* (1851) and Joris-Karl Huysmans's *À Rebours* (1884).[49] In the introduction to his English translation of *Captain Pamphile*, Andrew Brown describes the work as "an oddly disconcerting text."[50] For those who are used to thinking of Dumas as the author of works like *The Count of Monte Cristo* (1844) and *The Three Musketeers* (1844), it is indeed jarring to read this book that seems, at first glance, like the far-fetched tales of a rapscallion but quickly reveals itself to be a jumbled narrative that interweaves anecdotes about a Parisian apartment filled with a long line of dying pets with the exploits of a racist sea captain.[51]

In *Captain Pamphile*, white settler colonialism and the enslavement of African people go hand in hand. Partway through the story, Pamphile, having been thrown overboard by his mutinous crew, arrives at the mouth of the St. Lawrence River, in what is now the unceded First Nations territory known as Quebec. There, he is taken on as a servant of the chief, Black Serpent, who is baffled by the mechanisms of Pamphile's wind-up watch yet wise in arts like navigation, a classic repetition of the noble savage trope. Pamphile later flees Black Serpent, disdaining "the tender care the great chief had taken of him," and regains control of his ship.[52] Pausing for various acts of treachery along the way, he eventually sails to the west coast of Africa, where he heads inland along the Orange River, which today makes up part of the border between Namibia and South Africa. Although he had intended to pick up a cargo load of ivory, he changes his mind and decides to take on a load of "ebony wood": 230 Africans, each of whom Pamphile allocates one square foot of physical space in the ship's hold with a cold, meticulous calculus.[53] After a voyage to Martinique that lasts two and a half months, Pamphile sells the enslaved people at a "premium price," reporting with a despicable businessman's pride that he had "lost only thirty-two Negroes."[54]

How do we make sense of this vile character as the prototype for the sailor who is imagined to have used the dames de voyage? The answer is complicated, in part because *Captain Pamphile* is complicated. As a character, Pamphile's racism is undeniable and unforgivable. Yet the racial politics of the novel are harder to pin down. Dumas was himself of mixed race;

born in France in 1802 to a white mother, he was the grandson of a formerly enslaved Haitian woman on his father's side.[55] Thus, *Captain Pamphile* is a novel about a man who enacts the horrifyingly callous treatment of enslaved African people as he transports them to sell in the Caribbean, written by a man whose own family lineage includes people from the African diaspora enslaved and sold in a very similar manner. Certainly, Pamphile is far from a sympathetic character, and we are clearly not meant to feel good about his decision to swap ivory trading for slave trading. Yet there is also no comeuppance for Pamphile. In some other story, slave trading might be the final straw, the point at which a character's actions (among them, in Pamphile's case, cruelty to animals, cruelty to crew members, and cruelty to Indigenous peoples) have finally gone too far. As for Pamphile, he simply turns a handsome profit and moves on to his next heinous endeavor. From one perspective, this is a biting portrait of a slave trader who is himself far less than human; from another, Dumas neither condones nor condemns Pamphile, allowing him to remain a worrisomely laughable scoundrel.

In fact, it is that very ambiguity that makes Pamphile a fitting symbol of the racial dynamics that underlie the tale of the dames de voyage. Later authors seem to remember him as a stand-in for a generic lustful sea captain; rather than racist and cruel, the author of *La Femme endormie* describes him as desirous and "brave."[56] Yet in Dumas's novel itself, we find that the issues of power, colonialism, and race that others overlook—both in referencing *Captain Pamphile* and in telling stories about the history of sexual technologies—are in fact palpably present. Read as a minor yet nonetheless revealing part of the cultural history of the dames de voyage, Dumas's novel makes explicit the resonances between the story of the sailors' dolls and the very real histories of the dehumanization of racialized subjects. It reminds us that the cultural imaginaries that surround technology cannot be separated from the cultural imaginaries that surround race. What's more, it suggests that the true origins of what we now call *sex tech* may lie not in the creation of sex toys but in the establishment of a very different set of "technologies": the transnational systems of governance and economy that made colonialism and slavery possible on a grand scale.

Seen in this manner, the links between *Captain Pamphile* and the imagined origins of the sex doll can help us identify what we might understand as a prehistory of the contemporary racial dynamics that surround newer sexual technologies like sex robots and related forms of artificial intelligence. Writing in their book *Surrogate Humanity*, Neda Atanasosko and Kalindi Vora draw parallels among what they refer to as twenty-first-century "design imaginaries of sex robotics," the capture of African people, and histories of US racial slavery.[57] Atanasosko and Vora are careful to explain that they are not drawing a direct equivalency between enslaved people and "robot nonsubjects"—a clarification that I too echo.[58] However, like Atanasosko and Vora, I am interested in the imaginaries that circulate around and come to shape sex tech (in this case, imaginaries of history rather than imaginaries of design) and how they have themselves been shaped by unspoken yet nonetheless foundational notions about race and who counts as human. Long before the rise of sex robots, we find that their predecessors, sex dolls, were already being imagined through their implicit relationship to a history of slavery and capture that, as Atanasosko and Vora say, has for centuries assigned "humanity along racialized and gender hierarchies."[59]

Reimagining History beneath the Water

Even when we draw out the alarming echoes between the tale of the dames de voyage and histories of colonial and racial oppression, there remains a potentially powerful opportunity to find new meaning in the story of the sailors' dolls. As Nettrice R. Gaskins has explained, it is critical to simultaneously address issues of racial discrimination that pervade technology across its history and to push for new ways of understanding the relationship between technology, people of color, and subjects of the Global South.[60] "Historically, people of color have been casualties of technologically enabled systems of oppression," Gaskins writes.[61] However, by looking at how factors like "the ways that ethnic communities of practice have voluntarily subverted or remixed dominant technologies," scholars can move beyond simplified notions "that people of color are . . . forever victimized by the outcome of [technology's] production."[62] Another approach to shifting these dominant narratives is to turn to the work of what Ruha Benjamin calls the *liberatory imagination*: speculative thinking about technology that "opens up new possibilities and pathways, creates new templates, and builds on a black radical tradition that has continually developed insights and strategies grounded in justice."[63] This liberatory imagination offers just such an opportunity to develop new insights into sex tech's history through acts of speculation.

I am thinking here specifically of the ways that the tale of the dames de voyage resonates with Black feminist and Afrofuturist reimaginings of the Middle Passage. The image of the sex doll thrown overboard, which is described glibly in a handful of the contemporary texts that recount the story of the sailors' dolls, suggests connections between the imagined histories of sex tech and the real histories of those cast overboard on slave trading ships, leaving behind what Mustakeem describes as a trail of the dead in "watery graves."[64] This historical practice—which, as Jessica Marie Johnson has demonstrated, can be seen as forming a predigital backdrop to present-day fixations on data—is abhorrent, yet Black authors and creators themselves have taken up this facet of history and reenvisioned it through the liberatory imagination.[65]

For example, rap artist Lupe Fiasco's 2018 concept album *Drogas Wave* tells the story of "a fictional group of Middle Passage slaves who survived drowning in the ocean and dedicated their lives to sinking slave ships from under the water."[66] Relatedly, the 1997 album *The Quest*, created by electronic music group Drexciya, "[establishes] an origin myth based on the Middle Passage." As Gaskin explains, "The world of Drexciya begins with the creation of an underwater country populated by the unborn children of pregnant African women thrown off of slave ships," bringing issues of gender and sexuality to the fore and demonstrating how the descendants of survivors of the Middle Passage transformed the Atlantic Ocean into a "place for artists to imagine alternate worlds and realities."[67] Both albums use techniques of Afrofuturist worldmaking, which Ytasha L. Womack describes as standing at the "intersection of imagination, technology, . . . and liberation," to transform histories of violence into visions of new life.[68] They serve as reminders that Afrofuturism is not just about the future, but also, at times, "a total reenvisioning of the past."[69] Artist Ellen Gallagher's ongoing *Watery Ecstatic* series, started in 2001, approaches this subject through visual interpretation, interweaving imagery of sea creatures and human bodies. As Suzanna Chan writes, Gallagher's pieces "feature the black Atlantic in countermemories that reinscribe the historical murder of African women through a myth of their survival and transformation into aquatic beings. The artworks defy contemporary eliminations of, and assaults on, black lives to claim a spectacular present and posthuman future."[70]

If the tale of the dames de voyage, as the story of the dolls passed among white European sailors and used and discarded at sea, contains echoes of the Middle Passage, might it too be fodder for a reenvisioning of the past? In the conclusion to this book, I argue for the possibility of reclaiming the dames de voyage: finding within the story of these dolls a queerer and more feminist way to imagine the origins of sexual technologies. The connections between the tale of the dames de voyage and real histories of colonialism and racism simultaneously complicate and animate that call to reinterpret the sailor's doll. One the one hand, any un-self-critical attempt to reclaim a figure, imagined or otherwise, that emerges from and indeed often perpetuates colonialist and racist cultural logics would be deeply problematic. Yet there is also much inspiration from Black feminist think-ers in lingering with the power of such an act of speculation, which does not change or fix history, but instead gives us new ways of making mean-ing from the past. In her book *Undrowned*, Alexis Pauline Gumbs describes how reflecting on the experience of the Middle Passage is, for her, also a way to reflect on the act of breathing: in the past and in the present, as a metaphor and as an embodied reality.[71] "Breathing in unbreathable cir-cumstances," writes Gumbs, "is what we do every day in the chokehold of racial gendered ableist capitalism."[72] Thinking alongside this history sparks new ways of thinking about the present and the future—whether it is the future of interspecies life on our planet and racial justice, as Gumbs explores, or the future of sexual technologies.

7 Legitimizing Sex with Technology: Prisoners, Nazis, Misogynists, and the Origin Stories That Go Untold

Prison cells, Nazi war camps, the bedroom of a mythical sculptor. These are some of the sites of alternative origin stories—most just as imaginary as the story of the sailors' dolls—for the birth of contemporary sexual technologies. In this final chapter, I transition from talking about the tale of the *dames de voyage* specifically to considering what other possible origin stories are not being told when authors tell sex tech's history through the story of the sailors' dolls. As I have mentioned, there are many ways to tell the history of sexual technologies and the sex doll. Some such narratives offer valuable opportunities for resurfacing the contributions of those who have largely been left out of dominant versions of history. The alternate origin stories I discuss here, by contrast, are not any more "true" than the tale of the dames de voyage. They too are the stuff of fantasy, contortions of fact, and problematic cultural logics. Nonetheless, they offer valuable lenses through which to consider how the history of sexual technologies is being told. In particular, they draw attention to what dominant historical narratives choose *not* to say, often precisely because doing so would reveal uncomfortable truths about sex tech in the present moment. By allowing authors to sidestep these alternate origin stories, the tale of the dames de voyage helps legitimize twenty-first-century sexual technologies like the sex robot, while keeping other possible interpretations of sex tech at bay.

In the sections that follow, I talk specifically about three alternate origin stories for the sex doll and for sexual technologies by extension. The first situates the birth of the sex doll not aboard seafaring ships but in prison. As I explain, prisoners have been gradually written out of the sex doll's history as part of an ongoing effort to render sex dolls either more appealing or less "deviant." The second story is more disquieting, placing the birth of the sex doll in the hands of Nazi party leaders during World War II, who, so the myth goes, invented the blow-up doll for use by German troops. This story, I argue, has been set aside in favor of the tale of the dames de voyage because it risks bringing to the fore the underlying eugenicist and white supremacist overtones that characterize current conversations about "sexual robot futures." Last, I look at the myth of Pygmalion, which is frequently told alongside the tale of the dames de voyage. The myth is commonly deployed to position sex tech within a venerable classical tradition. Instead, I perform a close reading of Ovid's widely influential telling of the myth, demonstrating how Pygmalion's story itself undermines the aura of gravitas that it is meant to impart to sexual technologies. I finish this chapter by reflecting on what these alternative origin stories—the case of the prisoners' dolls, the Nazi doll, and the mythical "ivory girl" who comes to life—can reveal about sex tech cultures of the present moment and the people who participate in them.

When it comes to the sex doll, perceptions of legitimacy also go hand in hand with perceptions of the "natural." In her writing on technofetishism, Allison de Fren argues that the members of the online alt.sex.fetish.robots community she studied (a kind of precursor to a now much wider and more visible cultural interest in sex with robots) were less invested in "technology in general, or the artificial woman in particular, than in a strategy of denaturalization."[1] Through their attraction to robots, as well as mannequins, dolls, sculptures, and other "feminized objects," de Fren explains that these technofetishists sought to hack or deprogram societal assumptions about "natural" sexuality, resisting the essentialist expectation that human beings are attracted to other human beings on an alienable level.[2] By contrast, I argue here that the origin stories typically foregrounded in dominant narratives about sex tech's history are precisely those that serve to naturalize the concept of sex with technology. The tale of the dames de voyage reflects an approach to imagining that history that is concerned, first and foremost, with rendering the sexual devices of today, as well as the people who use them, "normal." Holding this tale against other origin stories for the sex doll that have been pushed to the background allows us to question that bid for normalcy and to lay bare the real cultural problems within sex tech that it tries to hide.

Illegitimate Inventors: Creating Sex Dolls in Prison

One possible alternate origin story for the sex doll locates the act of their invention in prison. As possible backdrops for the imagined birth of sex tech, prisons and sailing ships would seem to have a lot in common. Although the realities of both settings are far more complicated—and while the racial politics of mass incarceration in the United States should never be far from our minds when discussing prisons, even as imagined spaces— prisons and sailing ships share a set of traits as they are depicted in media and popular culture.[3] They are typically envisioned as all-male spaces (notwithstanding the recent interest in women's prisons sparked by the success of *Orange Is the New Black*), where burly men are forced to share tight quarters for long stretches of time. If, as the tale of the dames de voyage claims, sailors' frustrated desires in centuries past could inspire them to invent the very first sex doll, why could this same inspiration not be imagined to have struck prisoners rather than sailors?

The answer lies in the matter of legitimacy. Taken as contemporary cultural archetypes, sailors are perceived as "better" people than prisoners. Sailors, especially those from centuries past, are imagined as a rough-and-tumble sort who are nonetheless hearty, hardworking citizens. (We know from chapter 3 that this perception has itself shifted considerably over time, even just within the context of American popular culture.) Prisoners, on the other hand, are seen as dangerous and unfit to participate in society. They are also implicitly racialized, whereas sailors are envisioned as white by default. Cultural perceptions of historicity work against the prisoner as well. Whereas the sailor is viewed as a charming rogue from long ago, the prisoner is viewed as a menace of the present. The sexual conditions of sailing ships and prisons, as they appear in the cultural imaginary, themselves differ in important ways. Once they arrive in port, sailors are often associated with sex and queerness—yet this association often does not follow them onto their ships, as we saw in chapter 5. Prisons, on the other hand, are envisioned as more explicitly sexualized spaces, sites of sex and sexual assault between men. As John Mercer has explained, eroticized myths about prison have featured heavily in gay pornography, where they become sites for exploring non-normative sexual roles and identities.[4] Yet whether viewed through the lens of gay panic or the lens of queer desire, visions of sexuality in prison are bound up with ideas of deviance. Thus, to say that prisoners rather than sailors invented the sex doll would actively work against the (hegemonic, heterosexual, white) legitimacy that the tale of the dames de voyage attempts to establish. It would make sex dolls themselves seem deviant—or, rather, it would uncomfortably highlight the ways in which sex dolls are often already placed in this category.

In fact, the idea that prison might represent an alternate origin story for the sex doll is not a hypothetical one. If we look back at the actual history of sex dolls, as well as the lineage of texts that brings us the established version of that history in the present day, we find that prisoners have already played an important role in the development of sexual technologies, though their contributions have largely been left out of this history. The clearest illustration of this erasure of prisoners from sex tech's history lies in the supposed "photograph" of the dames de voyage that I discussed in chapter 1: the black-and-white image of what appears to be three sex dolls dressed in women's clothing (see figure 0.2). This image is commonly presented in twenty-first-century texts as visual evidence of the existence of the fabled sailors' dolls. Yet the image is not actually a photograph, nor does it depict sex dolls made by sailors. Instead, it is an artist's rendering of a sex doll (wearing three different outfits) fabricated by a prisoner, as we can tell from the earlier German-language captions that accompany the image. This distinction is itself important for uncovering the real history behind the dames de voyage, but far more telling for our purposes here is the process of how illustrations of a prisoner's sex doll from the start of the twentieth century came to be repeatedly misinterpreted as images of dolls made by sailors.

To illustrate, let me briefly recap how the image has moved through sources over the last two decades: Uses of the image today can largely be traced back to its appearance in David Levy's influential 2007 book *Love and Sex with Robots*, in which the placement of the image strongly suggested that it was a depiction of the sailors' sex dolls.[5] Levy received the image from Cynde Moya, who included it in her own dissertation, "Artificial Vaginas and Sex Dolls," with a note that she thought perhaps it represented the dames de voyage, though she was not sure.[6] Moya cited her source for the image as an illustrated volume that accompanied Leo Schidrowitz's 1927 German sexological text *Ergänzungswerk zur Sittengeschichte des Lasters* (Supplement to the moral history of vice).[7] Digging up Schridowitz's text and looking more closely at the image (which has since been replicated so many times in later printings that its details are hard to discern) clarified that the image includes two captions. The first, a more formal typeset caption beneath the image, translates to: "Coitus-substitute: undressable, primitive doll almost life-size, equipped with all female clothing." A second caption, this one handwritten on the image itself, translates to: "This doll was made by a prisoner in order to have coitus with her."[8]

There are a number of things we could note in this lineage—such as how "history" changes when materials are passed like whispers down the lane or how sexual history is remade through interpretation and imagination. However, the thing to notice for the moment is how the status of the doll specifically as a *prisoner's* doll was gradually erased over time. In the present day, we see authors of English- and French-language texts reprinting the supposed image of the dames de voyage while recodifying the assumption that the doll(s) represented can be attributed to sailors. Yet the work of separating the image of the doll(s) from its association with prisoners started much earlier. This is evidenced, for example, by the set of double captions in Schidrowitz's text. The caption that explains that the doll was created by a prisoner is handwritten and harder to read; it was perhaps written by an archivist at the Institut für Sexualwissenschaft (Institute of Sexology) in Berlin, to which Schidrowitz credits the image. Yet in his own, more formal caption, Schidrowitz has notably sidestepped any mention of a prisoner. Instead, he describes the doll only as a "coitus-substitute." Although the reason for this choice remains unclear, it seems driven by an erotic logic that is itself imbricated with a logic of legitimacy. In its own way, Schidrowitz's text is designed to be both scientific and titillating. Labeling the image as a *sex doll* rather than a *prisoner's sex doll* may have made it seem sexier.

The story of this image takes yet another twist if we trace it back to an earlier source. Seven years before the image of the prisoner's doll appeared in Schidrowitz's text, a version was published by the imminent sexologist Magnus Hirschfeld himself in a 1920 volume of his book *Sexualpathologie* (figure 7.1).[9] Situated within a discussion about people who are sexually attracted to visual representations and replicas of the body, Hirschfeld brings up the Greco-Roman myth of the sculptor Pygmalion, adding, "Primitive men and peoples themselves make much more primitive replicas than the artistic and skillful Helens. As an example, I offer pictures of a doll that a prisoner made as a substitute for a woman."[10] Here Hirschfeld prints

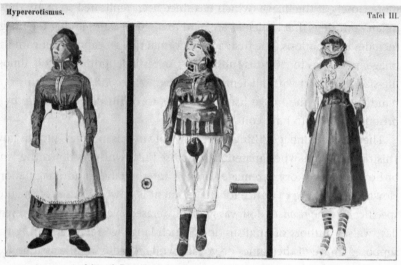

Figure 7.1
Three renderings of a life-sized sex doll made by a prisoner. This image (undated) appears in volume three of Magnus Hirschfeld's *Sexualpathologie*, published in 1920.

his own copy (seemingly more complete and closer to the original artwork) of what we now know to be the image of the prisoner's doll.

For the first time, in Hirschfeld's text, we can see the image clearly. It is certainly not a photograph; the strokes and shading of what was likely an ink, watercolor, and/or gouache artist's rendering are very visible. Also visible is the intricacy of the doll. Her face looks carefully constructed and her body is more or less realistically proportioned, though she is conspicuously missing hands. Her wardrobe is impressive; it seems to include a dress, a skirt, socks, shoes, even a dainty scarf to tie around her neck. Even more informative, from the perspective of sex toy history, we can see that she has been equipped with a specific set of apparatuses for sexual penetration. To engage in sex with her, it seems, a person inserts their genitals into a tube with a soft, pseudovaginal opening, much like a modern-day Fleshlight. The tube is then inserted into an opening in the doll's pubic area. This opening itself appears to consist of a hole cut into the doll's undergarments, which has then been fitted with an inflatable rubber *ventre de femme* like those discussed in chapter 4. Thus, though this is not an image of a dame de voyage as it is imagined in the present day (as a sex doll made by sailors), it is in fact an image of a sex doll wearing an artificial vagina of the sort that was sometimes referred to, in the original historical context, as a *dame de voyage*. Things have come full circle.

Unlike authors after him, Hirschfeld is not shy about attributing the creation of this sex doll to a prisoner. In addition to the description he provides in the body of his text, he also captions the image: "Life-size doll made by a prisoner as a substitute for a woman."[11] This does not mean, however, that his recognition of the prisoner's creation is complimentary. He uses this image to illustrate that "primitive" people like prisoners (the word *primitive* bringing with it both racial and colonial overtones) make sex dolls that are supposedly less artful than those made by the "Helens"—that is, figures from Greek and Roman antiquity. Interestingly, in this juxtaposition, prisoners are not presented as illegitimate because of their association with social deviance. Instead, what makes prisoners illegitimate in Hirschfeld's description is their purported lack of artistic skill. This is an odd choice, given that the doll in question seems to be quite artfully made. It also contrasts sharply with the contemporary tale of the dames de voyage, which celebrates sailors as ingenious inventors precisely for fashioning dolls that are often described as rudimentary. In this sense, the case of the prisoner's doll illustrates how legitimacy and innovation, as they have been posited across various histories of sexual technologies, are tied more to who does the inventing than to the quality of what is invented.

Despite prisoners' erasure from sex tech history, it is actually in the inventions of prisoners that we find real makeshift sex dolls that most closely resemble the imagined sailors' dolls. Some of these dolls do not date from times long past but from the twenty-first century. Written by author Angelo, the 2003 book *Prisoners' Inventions* catalogs an array of clever contraband devices that Angelo and his fellow prisoners have made while incarcerated.[12] A number of the devices that Angelo describes are related to sex, such as a "covered wagon" (a blanket fort designed to give privacy for sex and masturbation) and a variety of prophylactics, including a condom made out of sandwich wrap and a rubber band that Angelo playfully refers to as a "French tickler."[13]

Of particular note is an entry in *Prisoners' Inventions* labeled "muff bag." The muff bag replicates a pair of buttocks with a torso and legs that can be penetrated for simulated sex. Angelo explains how the item is constructed from water-filled trash bags and a set of blankets: "The device is made using two 24-inch square thin plastic wastebasket liners (a double-bag system is recommended due to the tendency of the bags to puncture in moments of passion). The corners of the bags are pulled in and tied together inside the bag so that the whole thing resembles Speedo shorts without the leg holes, after which the bag is filled with hot water, tied off, and positioned properly on the bunk. . . . Add rolled-up blankets to simulate the torso and legs and supply support and you're ready to slip it to her."[14] This description, along with Angelo's accompanying illustration, paints a picture of a fornicatory doll much like the envisioned dames de voyage, made in a human form out of materials ready to hand. Yet the muff bag also productively complicates the vision of the sailors' dolls by inhabiting an ambivalent space between variously gendered bodies. The item is referred to with the slang term *muff* but Angelo says that the bag, once tied up, looks like men's Speedo shorts. Angelo's illustration resembles the torso and legs of a cisgender man, yet his description of the item concludes, "You're ready to slip it to her." Ultimately, it remains unclear whether the muff bag is supposed to replicate an ass or a vagina, or whether perhaps the item itself blurs the division between the two.

The depictions of the prisoners' sex dolls that we see in Angelo's texts differ in critical ways from Hirschfeld's. Angelo himself explicitly situates the creation of prisoners' inventions within traditions of technological innovation. For example, he describes one particularly inventive cellmate as "possessing the spirit of Thomas Edison," calling him a veritable "Leonardo da Vinci of his age."[15] Angelo is also careful to explain that this "technology" of the prisoners' inventions, as he refers to it, is all the more impressive given severely limited access to materials and oppressive regulations in prison. "The prison environment is designed and administered for the purpose of suppressing such inventiveness," he writes. "Officially, the devices described here are considered contraband, subject to confiscation in routine cell searches. But inmates are resilient if nothing else—what's taken today will be remade tomorrow."[16] In these ways, the image of the prisoners' doll (as an actual present-day technology) destabilizes the image of the sailors' doll (as an imagined historical technology). It places sexual inventiveness in the hands of people whom mainstream society sees as deviant. At the same time, it flips the script on that notion of deviance, recasting prisoners like those whose inventions Angelo documents as bold inventors, ones who refuse to allow the oppressive regimes of prison to prevent them from engaging in sexual expression. Therefore, as an alternative point of origin for contemporary sex tech, the prisoner's sex doll sheds light on dominant, hierarchical notions about who gets to be an inventor and who, to call back to Hirschfeld's language, is deemed an "unartful primitive."

Nazi Blow-up Dolls: An Urban Legend

While some alternate origin stories for sex dolls risk being seen as delegitimizing, others are outright damning. One such story is the tale of how sex dolls—and blow-up dolls in particular—were supposedly invented by Nazis for use by German soldiers during World War II. Like so many stories about the origins of sexual technologies, the specifics of this myth, as it appears in contemporary writing, vary from source to source. A few of these sources are among those discussed elsewhere in this book. For example, Rebecca Clark writes, "A popular urban legend holds that during WWII Nazi Germany engineered the first modern sex dolls—perfect Aryan specimens . . . in order to both train and contain the sexual appetites of its invading armies."[17] Clark points to Anthony Ferguson's *The Sex Doll* as her source for this anecdote.[18] In a chapter subsection titled "The Borghild Project," Ferguson gives an expanded version of the story, explaining that the Germans "supposedly created a special task force" to address the problem of soldiers' sexual frustration while away from home:

> [They] called it the "Model Borghild" project. Legend has it that the Nazis under Heinrich Himmler started the project in the early years of the war to combat the sexual excesses of the conquering German armies. Himmler was allegedly concerned about the debilitating effect of sexually transmitted disease picked up by his troops from foreign prostitutes of inferior races. His solution was to commission a select team of experts to create a traveling army of gynoids to follow the conquering Wehrmacht across the battlefields of Europe. The dolls were to be housed inside a series of "disinfections-chambers."[19]

Accounts from additional sources contrast with Ferguson's, claiming that it was not Heimrich Himmler, a Nazi party leader often described as "the architect of the Holocaust," but Adolf Hitler himself who provided the impetus that kicked off the Borghild Project.[20] Some accounts also elaborate on the specific concerns of Nazi leadership that lead to the creation of the dolls—for example, that German soldiers would contract syphilis from Parisian sex workers.[21] Clark describes the dolls as "perfect Aryan specimens," but the story as it appears across a broader network of sources is more complicated. Many authors imply that the prototypical doll in the Borghild Project was indeed white, blue-eyed, and blond, but clarify that she was actually intentionally designed *not* to look like a "good" German wife. Instead, reportedly, she was given a "coquettish" face so that she would have the air of a sex worker.[22] This reminds us, yet again, of the complicated yet notable role that sex workers have played in the history of sexual technologies, whether as creators and users of such technologies or as figures of fantasy that have driven sexual imaginaries.

There are a number of similarities between the story of the sailors' sex dolls and the myth of the Nazi blow-up dolls. For instance, both origin stories are, in reality, not actual, factual histories but instead comparatively recent constructions. As we know from chapter 4, blow-up dolls (initially made of inflatable vulcanized rubber) may have been manufactured in Europe or America as early as the 1850s, with commercial versions of such dolls on the market by the 1880s or 1890s. Just as sailors did not invent and popularize the sex doll, as the tale of the dames de voyage would have us believe, Nazis did not invent the blow-up doll. The two stories also come into being in similar ways. Although my investigation of the citational history behind the evolution of the dames de voyage has been far more extensive than my work on the story of the Nazi blow-up doll, initial research suggests that the two stories coevolved through related processes of hearsay and fantasy, with historical sources of information about the Borghild Project, like those of the tale of the dames de voyage, revealing themselves under scrutiny to be questionable at best.[23]

Yet there are also revealing differences between the two stories, both in their contents and in the ways that they are told. The myth that Nazis invented the blow-up doll positions sex dolls as military technologies—as opposed to technologies of pleasure or even technologies of colonialism, as we see in the tale of the dames de voyage. This is in keeping with the actual fact that many domains of technological development, as numerous scholars have noted, have emerged in conjunction with wartime efforts and military funding.[24] Even though it is false, the myth of the Nazi blow-up doll helpfully reminds us that the connection between technology and war extends to sexual technologies. Yet interestingly, the myth that Nazis invented the sex doll is comparatively infrequently told in the context of sexual histories that foreground technology, with some exceptions. Instead, it more commonly appears in histories that are intended to be quirkier or more comedic in nature, presenting sex dolls as curiosities with an even more curious history. (Personally, as a Jewish American scholar, I cannot say that I find the notion that "the architect of the Holocaust" invented sex dolls to be very funny, even if Nazis themselves are meant to be the butt of the joke.)

Another key difference lies in how the two stories are represented as fact versus fiction. Whereas the tale of the dames de voyage is repeated time and time again as being simply true, the myth of the Nazi blow-up dolls is typically discussed as an "urban legend" or an "unsubstantiated theory."[25] For every article or blog post that claims that the Nazis did indeed invent the sex doll, there is another that explicitly questions that claim. Consider this string of nearly identical headlines: "Did Hitler Invent the Inflatable Sex Doll?" "Did Adolf Hitler Really Invent the Sex Doll?" "Fact or Fiction? Hitler Invented the Inflatable Sex Doll . . . Allegedly!"[26] The difference in how the story of the sailors' dolls and the story of the Nazis' dolls is presented raises a question: Why is one origin story for the sex doll accepted as fact while another is presumed to be fiction? Perhaps the idea that Nazis invented the sex doll seems so outlandish that it inherently raises skepticism, whereas the idea that sailors did so seems more plausible, just unassuming enough to be "real." Ironically, there *is* a very real connection between Nazis and the history of sex dolls. As I mentioned in chapter 2, Nazi forces destroyed the Institute of Sexology in 1933, including archives that may have contained vital documentation regarding the sex doll's earlier history.

Of course, the most striking (and upsetting) feature of the myth of the Nazi blow-up doll is its basic premise. The proposition that the origins of the sex doll might lie in a "traveling army of gynoids" created to bolster the efforts of the Germany army casts this history in a far more sinister light.[27] Ferguson writes that the dolls were supposedly designed to be housed in a "series of disinfection-chambers," a notion that hauntingly recalls the use of gas chambers. Authors who tell this story often explicitly link the creation of the blow-up doll to ideologies of white supremacy. The Nazi dolls, we read, were expressly created to keep German soldiers from engaging in sex with "foreign prostitutes of inferior races."[28] As imagined, then, the dolls' role was to uphold a white sexual purity within a racist, xenophobic hierarchical system that positions sex with dolls as preferable to sex with nonwhite partners.

Why does the story of the Nazi blow-up doll not get told in the same contexts and with the same frequency as the tale of the dames de voyage? It's perhaps unsurprising that a story linking sex dolls to the Nazi party would be less popular among those who are themselves invested in presenting sexual technologies in a positive light. Yet at issue here is not just avoiding an origin story that would make sex tech look bad. The trouble with the story of the Nazi blow-up doll is that it draws too much attention to things that are already true about the dominant rhetorics surrounding sex tech today, especially when it comes to sex robots. As I explained in the introduction, the current buzz about sex robots is often animated by a set of romanticized, technoutopian visions of how such robots (almost always designed to resemble white or Asian women) will usher in a brave new future of mind-blowing sex.[29] Despite the intentions of any individual roboticist or proponent of robot sex, the very notion of building an ideal sexual partner cannot be separated from the logics of eugenics. The drive to create perfect people, whether as citizens or as sexual partners, is bound up with historical acts of violence precisely like those in Nazi Germany. Sex robots imply a hierarchy of racial and ethnic purity all their own, while allowing their creators to supposedly sidestep issues of race and ethnicity by moving beyond the human. This is the uncomfortable truth that the story of the Nazi blow-up doll, though it is itself a false history, makes all too apparent.

The Myth of Pygmalion: Making Masculinity Legitimate, Making Masculinity Absurd

Both prison cells and Nazi war camps may be unmentionable sites for the possible origins of today's sexual technologies, but there is one origin story that no one seems to hesitate to tell: the myth of Pygmalion. There are many versions of the myth of Pygmalion, which has been told and retold in numerous forms. Broadly speaking, as it's described in texts related to the history of sex tech, the myth is about a talented sculptor who creates a statue of a woman so beautiful he falls in love with it. His creation is then brought to life, and he marries her. Referring to Pygmalion, David Levy writes in *Love and Sex with Robots*, "Sex with humanlike artifacts is by no means a twenty-first-century concept—in fact, its foundations lie in the myths of ancient Greece."[30] Presented in this way, the myth of Pygmalion ties the sexual technologies of today to a time so long ago that it makes sex tech seem timeless. It is easy to see why the myth of Pygmalion serves as an appealing parable for the contemporary creation and use of sex dolls, at least on its surface. Pygmalion is an artisan, a man in touch with the gods, a symbol of antiquity that brings along with it all the cultural capital of Western classicism. Across these texts, Pygmalion comes to stand in for the figure of the (male) inventor who, through his vision and prowess, creates new works of technology so bewitching that even he cannot help but desire to possess them. He is seen as both legitimate and legitimizing: a legendary innovator whom every robotics engineer working on the newest sex doll AI could fancy themselves to be.

And, indeed, Pygmalion is everywhere. In the network of contemporary texts described in this book, references to Pygmalion proliferate. It's very common to see authors who recite the tale of the dames de voyage preface it by recounting the myth of Pygmalion, implying that while sex tech has a "real" historical origin story (the tale of the dames de voyage), it also has an origin so ancient that it goes beyond history (as modeled through the myth of Pygmalion). This is true in both twenty-first-century works and ones from earlier, back across the twentieth century and into the later decades of the nineteenth. Pygmalion gets a reference in Auguste Villiers de l'Isle-Adam's 1886 L'Ève future, Madame B.'s 1899 La Femme endormie, and René Schwaeblé's 1904 Les Détraquées de Paris.[31] A discussion of Pygmalion prefaces an explanation of the dames de voyage in Iwan Bloch's much-quoted 1907 The Sexual Life of Our Time.[32] Admittedly, rather than Pygmalion himself, Bloch is more interested in the fetish he refers to as pygmalionism, which he says entails enjoining a woman to take off her clothes, stand on a pedestal like a statue, and "gradually come to life"—a practice that Bloch compares to necrophilia and which, he says in the most unflattering terms, largely appeals to "old, outworn debauchees."[33] Recent accounts in this lineage offer a far more romanticized portrait of Pygmalion. In The Sex Doll: A History, Anthony Ferguson writes that "the origins of man's desire for the perfect love object date back to the ancient world," as evidenced by Pygmalion, who was "able to consummate his love for this perfect object of desire made flesh."[34]

If we look at the actual myth of Pygmalion itself, however, we find a story that does not match up with this idealized vision of perfect love objects or the venerable history of men's desires. Consider the version of the Pygmalion myth that appears in Ovid's *Metamorphoses*: arguably the most widely influential version of the tale.[35] In Ovid's "The Story of Pygmalion," Pygmalion is not so much a brilliant sculptor—the protofigure for the contemporary sex tech genius—as he is a timid, self-involved, misogynistic loner. At the story's start, Ovid writes that Pygmalion has sworn off women because he finds their behavior shameful. "Shocked at the vices nature had given the female disposition," Pygmalion chooses "to have no woman in his bed."[36] Instead, he makes for himself "with marvelous art, an ivory statue, as white as snow, and [gives] it greater beauty than any girl could have," at which point he promptly falls in love not with the statue itself but with his own "workmanship" (*operis* in Latin, arguably closer to "work" or "creation").[37] He touches and kisses the statue, then compliments her, and then starts bringing her an array of "presents such as girls love": trinkets ranging from flowers and jewelry to dresses and even "little pet birds," all of which he drapes across her ivory body in an act of what might best be described as girlfriend dress-up.[38] Eventually, though, he decides she looks better naked, removes the items, lays her down on a blanket, and "takes her to bed"—the same bed where he chose not to take real women because their morals were so shameful, whereas his naked ivory statue somehow maintains the air of a snow-white "virgin."[39]

As for the transformation of Pygmalion's statue into a flesh-and-blood woman, Ovid does not actually credit it to the sculptor's exceptional artisanship, but rather to an act of a goddess who takes pity on his romantic plight. Ovid's telling has Pygmalion attend a festival for Venus, whom he sheepishly asks for a wife just like his "ivory girl."[40] Pygmalion does not really want a wife, Ovid explains; he is just too embarrassed to admit that what he would actually like is to marry the statue he's already been sleeping with. Venus gets the point anyway (she "[understands] the prayer's intention"), and when Pygmalion goes home and starts kissing and stroking his statue, the statue seems to come to life—or, at least, it starts to seem vaguely alive.[41] We read how, beneath Pygmalion's fingers, the ivory of the statue softened like warm wax "made pliable by handling."[42] Pygmalion then has sex repeatedly with his "pliable" but still apparently silent partner. In the heat of his passion, he takes a moment to observe that the statue beside him has become a body ("*Corpus erat!*" writes Ovid [It is a body!]), with blood pulsing in its veins, blushing cheeks, and eyelids that open.[43] However, even now that she is living, the statue-turned-woman seems little more than a doll. She does not seem to speak or move her body of her own accord. With her wax-like skin, which Pygmalion touches "over and over," she is still basically an object. The one thing she can do, however, is get pregnant. Writes Ovid, "The crescent moon fills to full orb, nine times, and wanes again, and then a daughter is born, a girl named Paphos, from whom the island [on which the story takes place] later takes its name."[44] From here we hear no more of the ivory girl or her life as the "wife" of Pygmalion. For all the fuss around Pygmalion's statue coming to life, neither Pygmalion nor Ovid seem particularly interested in the life she leads.

This fact is not lost on Ovid, and indeed his rendition of the story of Pygmalion is itself highly critical of the sculptor. In Ovid's telling, Pygmalion is a comedic character: immature, misguided, and foolishly idealistic. Ovid's long description of Pygmalion's courtship of his statue is clearly meant to be funny. The absurdity of their romance is driven home in one particularly parodical scene in which Pygmalion lays the nude statue on a "crimson coverlet" with a "soft pillow under head, as if she felt it," and calls out to her with all campy sincerity, "Darling, my darling love!"[45] The arrival of Pygmalion's daughter, Paphos, at the end of the story also casts the tale in a more ambiguous light. On the one hand, this ending invokes cycles of abuse; one fears what might happen to a girl in the home of a father who hates women and so married his sex doll. On the other hand, there is potential for a kind of retribution in this ending. The man who found women despicable ultimately has his own story hijacked, in its final moments, by his daughter. She will go on to become the namesake of an island. Meanwhile, Pygmalion's name will become, among other things, a term for obscure sexual proclivities that German sexologists will later attribute to "old, outworn debauchees."[46] Thus, we find that even as the myth of Pygmalion is used to legitimize the practice of men making—and loving and sleeping with—dolls, Pygmalion's story itself makes the origins of sex tech seem far from legitimate.

Precisely because it fails to fit the legitimizing narrative that it is meant to serve, the Pygmalion myth helpfully draws out certain questionable elements of the imaginaries that surround contemporary sexual technologies. As an imagined origin (literally the stuff of myth), the story of Pygmalion suggests that what has sparked the invention of the sex doll and other such devices is not male ingenuity or male artisanship but rather a fragile, bitter masculinity. In Ovid's particular vision of Pygmalion, we see how a paradoxical, incel-like disdain fuels the creation of the sculptor's "ivory girl." Pygmalion reviles women because he thinks they are promiscuous and therefore sculpts himself a woman who is somehow both virginal and available for his sexual use at all times. He loves the statue, in large part, because it makes him feel powerful and talented; she puts up no resistance and reminds her creator-turned-husband, every time he looks at her, of his own exceptional cleverness. Like a misguided robotics engineer transported back to the time of gods and togas, he seems fascinated by the idea of transforming his statue into a real girl. Yet at the end of the day, the thing that really arouses him is how his creation teeters on the edge between real and unreal, human and nonhuman. She was so beautiful, Ovid writes, that she seemed "truly almost living" and yet her beauty was "greater . . . than any girl could have."[47]

What we find in the story of Pygmalion, then, rather than a legitimization or illegitimization of today's sexual technologies, is a reminder of what is really at stake in the telling of tales about the origins of sex dolls. References to the myth of Pygmalion attempt, and ultimately fail, to legitimize not sex dolls or sex robots themselves, but rather the people who make and use them. Such references reflect an effort to portray the particular brand of straight male desire that longs for sex with machines as a legitimate one. Whether we are talking about the myth of Pygmalion or the tale of the dames de voyage, both stories that are ostensibly about the creation of objects in the shape of women, what we are really talking about is men and how they want to be imagined. Here, we would do well to pause briefly and reflect on how Pygmalion's story contrasts with the story of Pinocchio, another work of fanciful fiction about a doll that comes to life, but one that has remained strikingly absent from cultural histories of sex dolls, automata, and other replica humans. Pinocchio's character originated in an 1883 Italian children's novel, but he is best known to contemporary audiences through Disney's *Pinocchio* film from 1940.[48] In this iconic representation, Pinocchio is a doll (well, a puppet) who longs to become a "real" boy. By contrast, in Pygmalion's story, the gender dynamics go the other way: Pygmalion is a man who longs to make a woman "real."

This intersection of gender and "realness" helps explain why it is Pygmalion and not Pinocchio we hear about when read about sex tech's origins. To suggest that men themselves might be dolls, and moreover to suggest that their status as "real" men might be in question, would come uncomfortably close to revealing the anxieties about masculine realness that already underlie attempts to legitimize sex dolls and sex tech more broadly. Real men make women; they are not made. Origin stories like the myth of Pygmalion and the tale of the dames de voyage also function to lend sex tech a sense of historical realness. Whether or not these stories are themselves "realistic," they cast an aura of authenticity around twenty-first-century sexual technologies and make them seem more "real" by giving them a past. Yet ironically, what we find in Ovid's story of Pygmalion is a mockery of the very idea of realness. Pygmalion is a lust-driven fool who either does not know or does not care what it would actually mean for a statue-turned-woman to seem real. Venus grants Pygmalion's unspoken wish and brings his beloved ivory girl to life in the way that one indulges the whims of a child who is playing make-believe. Yet Pygmalion is not a child; he is a man-child, and the more "real" his creation becomes, the more he seems disconnected from reality.

Alternative Origin Stories, Uncomfortable Truths

Attending to these alternative origin stories allows us to see what particular narratives the tale of the dames de voyage is deployed to promote—and also what uncomfortable truths about present-day cultures of sex tech it is meant to help sidestep. I have talked in previous chapters about how the story of the sailors' dolls works to make the history of sexual technologies masculine, straight, white, and Western. Yet even before all of those things, arguably the most immediate goal of the tale of the dames de voyage when it is deployed as an origin story is to make contemporary sex tech and the people who use it seem normal. Those people (most typically though not exclusively men) who celebrate the supposedly inevitable arrival of high-tech sex robots made to look like human women have a vested interest in dispelling the cultural stigmas that surround love and sex with machines: stigmas that they commonly blame for any critiques of questionable ethics. In the face of claims that erotic interactions with machines are "unnatural," the tale of the dames de voyage renders the invention of such machines preeminently natural, placing the early construction of makeshift sex dolls into the hands of normatively masculine and conveniently historically vague subjects. Certain accounts take this fantasy of the natural further, emphasizing the naturalness of the materials out of which the sailors' dolls were purportedly made: cloth, leather, straw. This brings a soft, human quality to the impression of sexual technologies as the stuff of cold, unnatural materials like metal and even computational code, justifying the high-tech by insisting on its roots in the low-tech.

By now, as we near this book's final pages, we know the truth about the dames de voyage. Far from representing a real set of historical practices, the tale of the dames de voyage reflects and embodies a series of efforts as well as a series of affects: evidence of intermingled desires and anxieties. We can understand the dames de voyage as one particularly telling example of how fantasy becomes fact and how the histories of both sexuality and technology take shape through selective forms of imagining and reimagining. Yet as much as the tale of the dames de voyage tells us about what visions of history are deemed desirable, it also tells us about what visions of history are not. In this sense, the story of the sailors' dolls operates in a defensive mode. The masculinity reflected in the tale of the dames de voyage is an anxious masculinity. In a cultural context in which sex dolls still often appear in mainstream media as the stuff of humor and men who have sex with them are seen as jokes, the tale is one among many tactics for insisting that sexual technologies and the people who use them are credible. Through the tale of the dames de voyage, the twenty-first-century tech enthusiast pining for a sex robot is transformed, in the cultural imaginary, from the stereotype of an emotionally stunted loner to the latest in a long lineage of strong, adventurous men exploring new sexual horizons, whether on the high seas or in their bedrooms.

What makes the particular alternate origin stories discussed in this chapter important to address is not some imperative to shift the imagined origins of sex tech from one tall tale to another. Rather, confronting the contrasts between which histories sex tech does and does not lay claim to reveals the constructedness of sex tech's history itself. These juxtapositions show us what the tale of the dames de voyage would not like us to see. They show us that making sex tech's history requires ignoring the contributions of those who are seen as "deviant," like prisoners, because sex tech itself teeters just on the brink of being deemed similarly deviant. They show us that the clear resonances between eugenicist thinking and the longing to build the "ideal" robotic sex partner hit a little too close to home for those invested in sex tech. And they show us that even when dominant narratives try to make sex tech's history seem so boldly masculine that it is mythical, the reality is that indulgent visions of dolls brought to life are often driven by toxic masculinity. Such masculinity may have a predecessor in the figure of Pygmalion, but it is very much a thing of the present. Without our intervention—and a push toward more liberatory forms of sexual technologies that break from these ways of thinking—it will continue to set the terms for the sex tech of the future.

Conclusion: Reclaiming the *Dames de Voyage*—The Feminist Potential of a Fictional Past

After this long road of investigating, debunking, and interrogating the story of the sailors' dolls—after all of the sources that crumble under closer inspection, all of the sexual histories left untold, all of the discriminatory implications of contemporary narratives about the origins of sexual technologies—it's hard to imagine sticking up for the *dames de voyage*. Indeed, at the center of this book has been the claim that the dames de voyage are not in fact the origin point of today's sexual technologies, as they are commonly envisioned, and that understanding the history of sex tech in ways that are both more accurate and more socially just requires deconstructing this tale, confronting its meaning and the kinds of cultural work it performs, and finding new origin stories. To put this point in the simplest terms, we might say that this project has been all about what is wrong with the dames de voyage and what that wrongness reveals. Clearly, it would seem, it is time to jettison the tale of dames de voyage, casting it overboard from the ship of both scholarly and popular histories of sexual technologies.

And yet, how I want to end is actually by making a case that might appear to run counter to (almost) everything that has come before—a case for reclaiming the dames de voyage. Yes, the tale of the dames de voyage is a false history that has emerged through problematic scholarly practices and the codification of fantasy into fact; yes, the tale of the dames de voyage presents a version of history that overwrites marginalized people and obscures actual histories, including histories of violence and oppression. Despite all this, there remains a glimmer of radical potential in the story of the sailors' dolls itself, a potential for an intersectional feminist reclamation inspired in part by the Afrofuturist work discussed in chapter 6, which uses speculative thinking about technology to remake both the past and the future. Throughout much of this book, I have argued for destabilizing dominant narratives about history (or those that serve the purposes of dominant groups) and going "spelunking" through the archives and genealogies of the past in order to institute counternarratives. However, as I demonstrate here in this conclusion, these counternarratives need not only be found by discarding the stories we are already being told. The tale of the dames de voyage shows us that we can also remake history from the inside. Like the ship lured toward the sirens' call, or like the teleological lineage of technology's history once it comes under feminist critique, the hegemonic cultural politics of the tale of the dames de voyage can be steered off course. The figure of the sailors' doll herself can be reimagined, rendered into a powerful symbol of subversive desire, if we put aside the work of figuring out what is "real" or "imaginary" and instead think about what is possible.

In what follows, I explore these possibilities through a work that is itself feminist and explicitly fictional: Anaïs Nin's erotic short story "Mathilde."[1] First published in 1977 but written during the 1940s, "Mathilde" includes an anecdote about a rubber woman used for sex and passed between sailors. Nin's story sparks a new way of seeing the tale of the dames de voyage, one in which the daydreams of women blur the divide between sexual subjects and sexual objects. For the protagonist of "Mathilde," who both luxuriates in the idea of being a rubber woman and laughs at the men who would use her, speculation transforms a sexist tale into an intimate daydream about being passed from bed to bed on a long voyage at sea. Although Nin's story, which has its own problematic racial and colonialist overtones, itself merits critique, it also serves as an invitation to reconsider the history of sexual technologies by inhabiting rather than dismissing fantasy. It prompts us to imagine what erotic appeal stories like the tale of the dames de voyage might have for those other than straight, white, cisgender men who idealize the prospect of sex with machines. "Mathilde" also brings new queer valences to the story of the sailors' dolls, challenging us to consider whether our technologies might have interior sexual lives all their own. Reclaiming the tale of the dames de voyage specifically for those whom this origin story and others like it have marginalized is a way to orient ourselves differently toward history, turning historical imaginaries into fodder for feminist imagination.

Feminist Fantasies: Becoming the *Dame de Voyage*

As I have mentioned, much has been written about appearances of women dolls and robots, typically created to serve as sexual or romantic companions for men, in literature and film. Less has been written about how women themselves—along with transgender, nonbinary, and queer people—have used narratives about so-called gynoid robots and dolls to explore issues of gender, sexuality, identity, and power. Examples of such texts, while largely absent from existing narratives about the history of the sex doll, are nonetheless important. For instance, Mary Shelley's *Frankenstein* can be seen as a gender-swapped and perhaps even gender-queered version of stories like the myth of Pygmalion or the tale of the dames de voyage.[2] In Shelley's book, a male inventor (himself the creation of a female author) builds a creature in the shape of a man and brings him to life. Later in the text, Victor Frankenstein also builds a bride for his monster, but destroys her before handing her over to his original creature. We might read this violent, desperate act as a kind of feminist refusal on Shelley's part to turn her narrative into yet another tale about a man making a woman for the pleasure of a man. Relatedly, Bliss Cua Lim describes how director Vera Chytilová, in her 1966 film *Daisies*, uses representations of women as dolls in the mode of a feminist allegory, caricaturing an "overtly patriarchal ideal of femininity" while highlighting in the figure of the doll a childlike "unmanageability [that] is accompanied by a disposition to resist."[3] A spark of this "disposition to resist" can in fact be found in various stories of the sex doll, if they are similarly read as feminist allegories, including the tale of the dames de voyage.

Julie Wosk, writing in her book *My Fair Ladies*, describes numerous works that illustrate what she refers to as "the woman artist as Pygmalion."[4] These range from *Frankenstein* to the far more contemporary work of artist Cindy Sherman, in which Sherman constructs and reconstructs herself through careful portraits—a mix of disparate, performative identities intentionally cobbled together, not unlike the mismatched body parts of Frankenstein's monster.[5] Wosk also provides examples of women installation artists, photographers, and filmmakers whose work challenges both earlier representations of dolls created by men, such as the work of midcentury surrealists and Dadaists, and more recent work by roboticists whose research has come to fuel the contemporary "scientific" push for the development of sex robots. One such artist is Heidi Kumao, whose 2005 installation *Misbehaving: Media Machines Act Out* Wosk describes in this way: "While male engineers in the 1990s and the early twenty-first century were constructing new compliant females—pretty, lifelike women robots that embodied their fantasies and desires—Kumao in her *Misbehaving* series used sets of mechanized, engineered female legs to create witty female 'robotic performers' that were rebellious and defiant."[6] Exemplified by Kumao's series is the fact that women have not only been the subjects of writing about robots and dolls; they have also been the creators of work that, in many instances, simultaneously explores and critiques the cultural alignment between the female body and its technological constructions.

Nin's short story "Mathilde" too can be interpreted as part of this constellation of feminist work. "Mathilde" is one among many pieces of erotic short fiction that Nin authored during her literary career, many of which were written for commission and later anthologized and published in the volumes *Delta of Venus* (1977) and *Little Birds* (1979).[7] While Nin herself remained ambivalent about her erotica, biographers and scholars of her work have since interpreted these short stories as examples of feminist pornography, combining explicit sexuality with an emphasis on the perspectives of women characters, many of them queer.[8] Using a narrative structure that is common in Nin's erotica, as well as in literary pornography more generally, "Mathilde" follows one particular character as she winds through various sexual encounters. Unlike the artworks described by Wosk, Nin's story does not reimagine a woman as the new Pygmalion. Rather, in "Mathilde," the woman is one and the same as the doll herself. "Mathilde" situates the doll in the realm of sexual fantasy rather than couching it within considerations of artistry or technological innovation, but the story has implications for the way we envision technology nonetheless. It wrests visions of the sex doll, and by extension all of the sexual technologies inextricably bound up with it, from the hands of straight men and repurposes it as a languid vision of indulgent female pleasure.

When Nin's story begins, Mathilde is a twenty-year-old hat maker working in a locale that has cropped up time and again throughout this book: Paris. Cultured, adventurous, charming, and quick to laugh, Mathilde is eager to explore the romantic opportunities the world has to offer. After hearing that Parisian women are "highly prized" in South America for their "expertness in matters of love," Mathilde decides to travel to Peru.[9] Despite bemoaning the fact that her voluptuous figure has always drawn crude attention from men (such as a gray-haired writer who seemed to be a poetic romantic until he proclaimed that the sight of Mathilde made him "stiff in [his] pants"), Mathilde quickly takes up sex work once she arrives in Lima.[10] She turns her new hat shop into a boudoir and hosts dreamy, opium-fueled orgies for Peruvian aristocrats and their friends. Nin gives long, voluptuous descriptions of Mathilde's group sexual encounters. "Mathilde would lie naked on the floor," Nin writes. "All the movements were slow. The three or four men lay back among the pillows. Lazily one finger would seek her sex, enter it, lie there between the lips of the vulva. . . . Another hand would seek it out too. . . . Then, for hours, they might lie still, dreaming."[11]

Mathilde is not a sex doll, but she has many doll-like qualities. She plays dress up with herself as one might with a doll. Writes Nin, "Mathilde developed a formula for acting life as a series of roles—that is, by saying to herself in the morning while brushing her blond hair, 'Today I want to become this or that person,' and then proceeding to be that person."[12] The role she plays in the orgies that take place in her Peruvian boudoir is similarly doll-like. Nin describes Mathilde being touched, kissed, pleasured, and penetrated for two or three days on end. Martinez, one of her regular customers, regards Mathilde's body and dreams of a grotesque, fleshy figure like a collection of mismatched doll parts and a "sex [that] was also mobile, moving like rubber, as if invisible hands stretched it"—recalling the rubber of the *femmes en caoutchouc*.[13] In the story's dramatic final sequence, Mathilde agrees to go to the private room of a participant in one of her orgies, Antonio, who forces her to take cocaine, rendering her into a kind of living sex doll. "They lay on the floor and she was taken with an overpowering numbness," Nin writes. "Antonio said to her, 'You feel dead, don't you?' . . . Antonio took [a] penknife and bent over Mathilde. She felt his penis inside of her."[14]

These scenes reflect the ambivalence of Mathilde's experiences in her role as doll. When she is passed around among the men who visit her boudoir, she describes feeling immense pleasure and satisfaction. Yet in this culminating sequence, we see what happens when doll-like-ness is forced upon her. She is rendered into a doll without her consent, and as a result ecstasy is transformed into powerlessness. Taken as a whole, "Mathilde" is far less a fantasy about what it would be like to *use* a woman as a sex doll and far more a fantasy about what it would be like to be a woman being *used* as a sex doll. Yet this fantasy is also inextricable from the nagging concerns of reality. Under the guise of erotica, "Mathilde" becomes a reflection on the irreconcilable tension between the sexual desire to be treated like an object and the understanding that when men sexually objectify women, it is rarely on women's own terms.

At the very heart of "Mathilde," presented as a reprieve among scenes of orgies and ejaculation, is a story within a story—one that will, by now, sound strikingly familiar. This is the story of a rubber woman, made and used by a group of sailors on a long sea voyage. After one of her escapades of group sex, while still under the effects of opium, Mathilde decides to masturbate in front of a mirror. Following a "wild orgasm" brought on by using her fingers to penetrate herself both vaginally and anally, "as she sometimes felt Martinez and a friend when they both caressed her at once," Mathilde muses in this way:[15]

> After seeing her movements in the mirror [Mathilde] understood the story told to her by a sailor—how the sailors on his ship had made a rubber woman for themselves to while away the time and satisfy the desires they felt during their six or seven months at sea. The woman had been beautifully made and gave them the perfect illusion. The sailors loved her. They took her to bed with them. She was made so that each aperture could satisfy them. . . . The sailors found her untiring and yielding—truly a marvelous companion. There were no jealousies, no fights between them, no possessiveness. The rubber woman was very much loved. But in spite of her innocence, her pliant good nature, her generosity, her silence, in spite of her faithfulness to her sailors, she gave them all syphilis.
>
> Mathilde laughed as she remembered the young Peruvian sailor who had told her this story, how he had described lying on her as if she were an air mattress, and how she made him bounce off of her sometimes by sheer resilience. Mathilde felt exactly like this rubber woman when she took opium. How pleasurable was the feeling of utter abandon! Her only occupation was to count the money that her friends had left her.[16]

Thus, in the midst of Nin's story, the narrative of "Mathilde" pauses so that we can learn the tale of a sex doll passed around between men on a sailing ship. Diegetically, the tale stands in for Mathilde's own daydreams in the wake of her orgasm. Yet the story also serves as a metaphor for Mathilde's experiences more broadly. She identifies with the rubber woman, but also she *is* the rubber woman: on a long voyage far from home, shared between men, "untiring and yielding." Elsewhere in the story, Nin suggests ties between Mathilde's body and the ocean itself. When she begins to masturbate in front of the mirror, she sees "the odorous moisture like the moisture of the sea shells" and declares, "So was Venus born of the sea."[17] To the extent that Mathilde can be read as a sex doll, she is one whose story is inextricably linked to the sea.

The Sex Doll Bounces Back

Despite all of the problems with the tale of the dames de voyage, I continue to feel much as I did when I started this research: the figure of the sex doll at sea intrigues me. This story within a story that Nin has written into "Mathilde," this tiny work of erotica about the rubber woman and her use among sailors, perfectly captures what I find so alluring about the tale of the dames de voyage, no matter how imaginary it might be. Reading Nin's "Mathilde" was one of my first encounters with the image of sailors' sex dolls. Long before I read example after example of twenty-first-century historical texts claiming that the origins of today's sexual technologies could be traced to the sexual frustrations of men on long sea voyages, I read "Mathilde." Therefore, always in the back of my mind, even as I have aimed to stringently critique the tale of the dames de voyage, has been the tickle of a reclamatory erotic potential suggested, succinctly yet richly, by Mathilde's fantasy of becoming the rubber woman. What draws me to Nin's version of the story of the sailors' sex dolls is its intimacy, its warmth, its strangeness. In the space of a few short paragraphs, it opens up so many possibilities for reenvisioning the sex doll and its role in the stories we tell about sexuality and technology.

Unlike the tale of the dames de voyage, which has formed around the fantasies and cultural agendas of straight men, this is a story about women and women's distinctly non-normative pleasures. Mathilde reflects on the tale in the luscious moments after her own orgasm, while she still sits "[watching] herself in the mirror . . . the honey shining, the whole sex and ass shining wet between [her] legs."[18] She is clear that she identifies with the rubber woman and that the idea of being a sex doll excites her: "How pleasurable was the feeling of utter abandon!"[19] The rubber doll in question, we are told, was designed so that all of her "apertures" could satisfy the sailors—a statement that echoes Mathilde's own double manual penetration of her vagina and anus. In this fantasy, then, Mathilde is both sex doll and sailor; all pleasure belongs to her. It is also important that the story ends with Mathilde celebrating her own occupation as a sex worker. In contrast to the many ways that sex workers have been erased from the history of sexual technologies, Nin's story within a story concludes with Mathilde reflecting happily that her only responsibilities were to have sex and "count the money that her friends had left her."[20]

In ways that we might not expect in an anecdote about a sex doll, the tale of the rubber woman contained within "Mathilde" is also a story about women's agency. The rubber woman is an object, but she is also an object with power over the sailors who made her. Although she may have been created merely to "while away the time and satisfy the [sailors'] desires," she commands real, deep feeling, inspiring love as well as a litany of accolades—which pepper this vignette like the tittering praises of enamored paramours who find her "beautiful," "perfect," "good-natured," "generous," and "faithful." For all her charming qualities and sexual pliancy, the rubber woman is far from simply sweet and obedient, however. Her gift to the love-struck sailors is syphilis. We could read this comment as a kind of a punchline: a bawdy joke about sexually transmitted diseases at the doll's expense. But it isn't really the rubber woman who is being laughed at here. Immediately after we learn that the doll gave the sailors syphilis, Mathilde lets out a chuckle, thinking about the Peruvian sailor who told her about the doll. The sailors romanticized their carefully crafted rubber woman, imagining her as innocent and faithful (somehow to all of them at once). In return, the doll mocks the absurdity of this fantasy by giving them not love but disease. Mathilde's own story ends on a sobering note that is far from funny. Yet in the rubber woman's story, women get the last laugh.

The way that men and masculinity are represented in this story of the rubber woman also differs strikingly from their role in the contemporary tale of the dames de voyage. While many of these recent, supposedly historical accounts explain the creation of the earliest sex dolls as the result of raw heterosexual desire and masculine ingenuity, here we see a tale that is much more tender, even feminizing. In "Mathilde," the sailors who create the rubber woman do beautiful work, creating the "perfect illusion." They feel warmly toward the doll and, what's more, their individual relationships with her actually foster among them a collaborative, mutually supportive community of men. The men on the ship in Nin's story seem to pass their rubber companion from bed to bed with magnanimity and kindness: "There were no jealousies, no fights between them, no possessiveness."[21]

Rather than using the figure of the sailors' doll to disavow queerness, as we see in the tale of the dames de voyage, the rubber doll in Nin's story expressly creates intimacy between men, becoming a vector of queer erotic relationality. The sailors in Nin's story are connected not only because they share the same lover, but also because their collective desire for the rubber woman allows them to share through mutual feelings. They may have sex with the rubber woman individually, but they love her together.

Nin's story of the rubber woman also compellingly scrambles the relationship between object and subject. In one sense, it is a story about women's pleasure, but it also decenters the human, shifting subjectivity and pleasure into the doll herself while insisting on her status as an object. Seen in this way, Nin's rubber woman takes on a queer, nonhuman animacy, bringing an explicitly sexual valence to visions of the nonhuman and posthuman theorized by scholars like Mel Chen, Anna Tsing, and Donna Haraway.[22] For at least two centuries, accounts of sex dolls (or what are functionally sex dolls) written by male authors and scholars have all shared a fascination with the idea that the sex doll might one day become so elaborate that it seems to be alive. We see this in contemporary efforts to make sex robots so realistic that they can pass for human women. It also manifests across the history of sexual technologies presented in this book— for example, in the recurring fascination with the idea that early rubber sex dolls might have been able to secrete replica vaginal fluids, coming to life, so to speak, through the sexual mechanisms of the body. By contrast, the rubber woman in Nin's "Mathilde" makes no attempt to come to life. She is not mechanized. She has no pneumatic tubes. No one tries to make her his wife. In fact, as her story progresses, Nin's rubber woman becomes increasingly less human. She begins as the "perfect illusion" but ends as a bouncy rubber mattress.

I talked at length in chapter 4 about the importance of rubber and its place within the material histories of sexual technologies. Here too, in "Mathilde," is it crucial that the doll that sails along with this lustful, loving crew is made of rubber. Yet rather than tying sexual technologies to issues of commercial production, as the real-life femmes en caoutchouc do, what makes the material qualities of the sex doll in Mathilde's fantasy so remarkable is its capacity to bounce. In this scene, rubber becomes a symbol of resistance. Nin writes of the Peruvian sailor who tells Mathilde about the rubber woman, "He had described lying on her as if she were an air mattress, and how she made him bounce off of her sometimes by sheer resilience."[23] This image draws out the physicality of the rubber woman, playing up both her object-like qualities and her unlikely agency. Whereas technologies of bounce have elsewhere been used to attempt to replicate and capture the movements of women's bodies, such as in the case of digital rendering software developed to create video games with supposedly realistic "breasts physics," in this tale it is the rubber doll's bounciness that makes her strong.[24] It allows her to bounce back both literally and figuratively. Pliant, she bends, only to snap back—using her capacity to bounce to send the would-be lover flying. Thus, she bounces back in the way that one fights back, simultaneously receiving lovers and fending them off, which she does with all the confidence of a nonhuman object who needs harbor no concern for men's fragile feelings. At the same time, the rubber woman also fucks back. She is not simply a sex toy or even a proto teledildonic device. Like Mathilde, the woman who dreams of being her, she is a sexual agent.

All of these qualities found within Nin's story also exist within the tale of the dames de voyage, if we know to look for them. The vision of the sailors' dolls cobbled together and shared at sea has emerged out of a long history of discriminatory thinking, and it has been put to many deleterious purposes. Yet it too contains possibilities for being reimagined as an intersectional feminist fantasy. For example, longing for the rudimentary sailors' doll—or even longing to be the sailors' doll—might offer an opportunity to resist the technoutopian and supposedly postracial fantasies that dominate current conversations about sexual technologies. It offers us an opportunity to undermine the fetishization of high tech, epitomized by sex robots, by directing our desires to low tech. For queer and transgender folks such as myself, the dames de voyage might also offer a point of nonheteronormative identification that draws from yet seeks to undermine dominant narratives about sex tech. The gendered valences of the sailors' dolls, like so many sex dolls, underscores the fundamental constructedness of gender itself. We might say that the sailors' doll is a cyborg figure, part human and part technology, but, inspired by work like Susan Stryker's, we might well also say that she is a trans figure.[25] She is a "woman" who has been made, not born. What she gives birth to, in this imagined origin story, is technology itself. These are just some possibilities for finding new meaning, and new pleasures, in the tale of the dames de voyage.

In some ways, it matters immensely that the true origins of sexual technologies, to the extent that they can be located anywhere at all, do not lie in the fictional account of the sailors' sex dolls. And yet in other ways, it does not matter at all. What matters is to recognize that when it comes to sexuality and technology, the line between history and fantasy has always been an illusion. Once we know that, the sexual technologies of the past and the sexual technologies of the future become ours to reclaim through the radical force of our own desires.

Sex Dolls at Sea

In the spirit of the long journey into a vast, watery expanse, I want to conclude by gesturing outward to an element of the tale of the dames de voyage that has floated in the background of this book but has not yet come to the fore: the sea itself. If the story of the sailors' dolls is a work of fiction that has been told and retold, in one form of another, for roughly the past 170 years, then the sea is a major character in this fiction. The sea is the setting for the supposed invention of sexual technologies: the vast yet vague (fittingly calling to mind the French *vague*, meaning "wave") locale, simultaneously geographic and mythical, where the first sex doll is envisioned as being born. In this vision, the sea is simultaneously a space of desire, of frustration, and of creation. As we saw in the first chapter of this book, various authors have fleshed out their accounts of the dames de voyage by adding colorful details about the dolls' construction or use. We could similarly expand on the tale by envisioning how the great, endless blue of the ocean might have stretched out before these sailors on their seemingly equally endless voyages, sparkling and undulating, inspiring an impossible longing for what lay beyond.

It is far from incidental that the tale of the dames de voyage places the origins of sexual technologies at sea. Despite the fact that this vision of sailors and their homemade sex dolls is meant to make today's sex tech seem more "natural," springing from a time and place unrelated to the computational technologies of the twenty-first century, the reality is that the ocean plays a major role in the infrastructures that maintain our contemporary digital world. We see this evidenced in work like Nicole Starosielski's research on the fiber-optic cables that cross the earth's oceans or Miriam Posner's research on the global supply chains that produce our electronic devices, which involve intricately interlocking systems of production and labor both "overseas" and across seas.[26] The ocean (and water more generally) is also one of our key structuring metaphors for making sense of technology and the flow of information, bringing us the growing prominence of terms like *streaming*.[27] Increasingly, with the rise of the "blue humanities," the ocean is becoming the realm of new scholarly inquiry, sparking calls to "think with water."[28] Turning to face the ocean is one of many ways to change our perspectives on technology and its history. Whether we are talking about cultural imaginaries or material realities, the ocean has never been a world apart from technology.

Positioning the origins of sex tech at sea, as the tale of the dames de voyage does, also has the potential to shift how we imagine technology's history in relation to gender. Contemporary histories of sex tech use the story of the sailors' dolls as part of a larger attempt to make the history of sexual technologies masculine, as I explained in chapter 5. Yet the sea brings with it feminine associations. The ocean has long been bound up with visions of sirens and mermaids, beautiful yet deadly. It is often seen as an enchanting but unforgiving expanse, at once maternal and capricious, as volatile in weather as women are imagined to be mercurial in mood.[29] Accounts of the tale tend to emphasize the sailing ship as a world populated entirely by men, with the sailors' doll seemingly the only "woman" for miles beyond measure. Yet around these men, in this imagined vision, lies the actual ocean, stretching out as far as the eye can see. The dame de voyage may be alone among sailors, in this scenario, but the sailors are themselves equally alone among the buffeting push and pull of the waves.

The sea also serves as a provocative space of possibility, a place for thinking differently about both history and media. "It offers a viable and transformative space of history," as Sowande' M. Mustakeem writes, a place from which to see the world in new, more fluid ways, from vantage points perhaps unavailable to us on solid ground.[30] Melody Jue, writing in *Wild Blue Media*, proposes an understanding of the ocean as a medium unto itself.[31] For Jue, seawater becomes a substance with the potential to defamiliarize "our terrestrial orientations."[32] Writes Jue, "The ocean is an environment for thought . . . a horizon of possibility, just beyond our experience but nonetheless part of the fabric and flesh of the world . . . unruly, testing the limitations of a human point of view."[33] The ocean then is the perfect place from which to reimagine the history of sexual technologies and our place within them.

Throughout this book, I have referred to the idea of sex dolls at sea. To be *at sea* is to be on the ocean, but it is also to be unmoored, adrift. Writing in the introduction to their special issue of *Women's Studies Quarterly* titled "At Sea," Terri Gordon-Zolov and Amy Sodaro explain that "the sea embodies dystopian despair, violence, and degradation, but also hope, coexistence, and possibility."[34] The notion of being at sea itself encapsulates these dichotomies, suggesting both what is lost and what might be found. For me, over the four years I have spent researching and writing this book, the sense that I am locating the history of sexual technologies at sea has taken on many meanings. There is the literal, of course: the tale of the dames de voyage, in its most common version, is a story about sex dolls made aboard ships. Yet in attempting first to track down and later to make sense of (what turned out to be) a false history, I myself have often felt at sea: drifting on a seemingly endless ocean of obfuscated sources, obscure archives, and disparate yet deeply interconnected areas of history and scholarship. Along the way, the material histories that undergird this project have served as a kind of life raft for me: a plank of wood to grab hold of amid the waves. And as the arguments and interventions of this book have come into focus, I have found that it is no longer me but instead the history of sexual technologies that is now at sea. Through this work, history has become unmoored. Hegemonic understandings of sexuality and technology are set adrift. This is as it should be. The tale of the dames de voyage now floats freely, carried off to new worlds of possibility by the waves.

Notes

Introduction

1. More of these news stories seem to come out every day. Here are just a couple of examples: Alex Williams, "Do You Take This Robot . . . ," *New York Times*, January 19, 2019, https://www.nytimes.com/2019/01/19/style/sex-robots.html; Paula Froelich, "Eerily Realistic Sex Doll Can Smile, Moan—and Even Hold a Conversation," *New York Post*, June 20, 2020, https://nypost.com/2020/06/20/realistic-sex-doll-can -smile-moan-and-even-hold-a-conversation/.

2. For more on Real Doll, see the 2012 documentary *The Mechanical Bride*, dir. Allison de Fren; Marquad Smith, *The Erotic Doll: A Modern Fetish* (New Haven, CT: Yale University Press, 2013); Tracy Clark-Flory, "What I Learned about Male Desire in a Sex Doll Factory," *Guardian*, October 19, 2020, https://www.theguardian.com/life andstyle/2020/oct/19/what-i-learned-about-male-desire-in-a-sex-doll-factory.

3. Some of the many possible examples include literature like writings by Gisèle Prassinos and Angela Carter, visual art like that by Hans Bellmer, television series like the anime *Chobits* (2000–2002), films like *Daisies* (1996), and much more. Gisèle Prassinos, *Arthritic Grasshopper: Collected Stories, 1934–1944*, trans. Henry Vale and Bonnie Ruberg (Cambridge, MA: Wakefield Press, 2017); Angela Carter, *Fireworks* (London: Virago, [1974] 2006; Sue Taylor, *Hans Bellmer: The Anatomy of Anxiety* (Cambridge, MA: MIT Press, 2000); Bliss Cua Lim, "Dolls in Fragments: *Daisies* as Feminist Allegory," *Camera Obscura* 16, no. 2 (2001): 37–77.

4. In addition to the books and articles about the history of the sex doll that I analyze in detail in chapter 1, some additional examples of recent work to reiterate this history include Kate Devlin, *Turned On: Science, Sex and Robots* (London: Bloomsbury Sigma, 2018), 38–41; Kate Lister, *A Curious History of Sex* (London: Unbound, 2020), 209–222.

5. Elen C. Carvalho Nascimento, "The 'Use' of Sex Robots: A Bioethical Issue," *Asian Bioethics Review* 10, no. 3 (2018): 231–240; John P. Sullins, "Robots, Love, and Sex:

The Ethics of Building a Love Machine," *IEEE Transactions on Affective Computing* 3, no. 4 (2012): 398–409.

6. Many scholars have addressed this issue of how logics of technological advancement contribute to discrimination. One particularly powerful articulation of this argument can be found in Ruha Benjamin's keynote address to the 2021 CHI conference: Ruha Benjamin, "Which Humans? Innovation, Equity, and Imagination in Human-Centered Design," ACM CHI Conference on Human Factors in Computing Systems, online event, May 11, 2021.

7. For more on methods related to Michel Foucault's concept of genealogy, as well as critiques of these methods, see Colin Koopman, *Genealogy as Critique: Foucault and the Problems of Modernity* (Bloomington: Indiana University Press, 2013).

8. Marshall McLuhan makes a similar argument when he writes in *The Mechanical Bride*, originally published in 1951, that mass media advertising reflects an "interfusion of sex and technology" that pervades culture, which McLuhan says is "born of a hungry curiosity to explore and enlarge the domain of sex by mechanical technique, on one hand, and, on the other, to possess machines in a sexually gratifying way." Marshall McLuhan, *The Mechanical Bride: Folklore of Industrial Man* (London: Duckworth Overlook, 2001), 106.

9. Shaka McGlotten, *Virtual Intimacies: Media, Affect, and Queer Sociality* (Albany: State University of New York Press, 2013); Paisley Gilmour, "Virtual Sex Parties—'I Just Attended My First Sex Party . . . on Zoom," *Cosmopolitan*, January 27, 2021, https://www.cosmopolitan.com/uk/love-sex/sex/a32266336/virtual-sex-parties/.

10. For more on internet-enabled sex toys, see Jeffrey Bardzell and Shaowen Bardzell, "Pleasure Is Your Birthright: Digitally Enabled Designer Sex Toys as a Case of Third-Wave HCI," in *Proceedings of the SIGCHI Conference on Human Factors in Computing Systems* (New York: ACM, 2014), 257–266.

11. Jonathan Coopersmith, "Pornography, Technology and Progress," *Icon* 4 (1998): 94–125.

12. For more on flirtations via telegraph, see Ella Cheever Thayer, *Wired Love: A Romance of Dots and Dashes* (New York: W. F. Johnston, 1880).

13. Regarding contraception as a technology, see Donna J. Drucker, *Contraception: A Concise History* (Cambridge, MA: MIT Press, 2020), 2.

14. For writing on teledildonics, see Howard Rheingold, "Teledildonics and Beyond," in *The Postmodern Presence: Readings on Postmodernism in American Culture and Society* (Lanham, MD: AltaMira Press, 2005), 274–287; Teddy Pozo, "Haptic Media: Sexuality, Gender, and Affect in Technology Culture, 1959–2015" (PhD diss., University of California, Santa Barbara, 2016); Maria Joao Faustino, "Rebooting an

Old Script by New Means: Teledildonics—the Technological Return to the 'Coital Imperative,'" *Sexuality & Culture* 22, no. 1 (2018): 243–257.

15. Ariane Cruz, "Techno-Kink: Fucking Machines and Gendered, Racialized Technologies of Desire," in *The Color of Kink: Black Women, BDSM, and Pornography* (New York: NYU Press, 2016), 169–212. The image of the "women's masturbation machine" pictured here comes from the visual supplement to Leo Schidrowitz's 1927 sexological text, *Ergänzungswerk zur Sittengeschichte des Lasters* (Supplement to the moral history of vice). Schidrowitz does not provide an exact date for its production but does include a caption that reads: "Commercially produced apparatus that was confiscated by the police and an original of which is in the Dresden Criminal Museum." Leo Schidrowitz, *Ergänzungswerk zur Sittengeschichte des Lasters* (Vienna: Verlag für Kulturforschung, 1927), no page numbers, figure V.

16. Jacob Kastrenakes, "Sex Toy Creator Finally Gets the CES Award She Was Denied," Verge, May 8, 2019, https://www.theverge.com/2019/5/8/18535907/ces -sex-toy-lora-dicarlo-award-reinstated-changes-promised. This claim that a sex toy could make its user "bawl with happiness" is a reference to an advertisement for a dildo in a catalog for a Parisian sex toy seller that was published in roughly 1908. It's part of the volume of collection catalogs currently held by the British Library that I discuss and provide more detailed citation information for in chapters 2 and 4.

17. Fred Turner, *From Counterculture to Cyberculture: Stewart Brand, the Whole Earth Network, and the Rise of Digital Utopianism* (Chicago: University of Chicago Press, 2008).

18. Regina Lynn, *Sexier Sex: Lessons from the Brave New Sexual Frontier* (Berkeley, CA: Seal Press, 2008).

19. Lynn Comella, "Studying Porn Cultures," *Porn Studies* 1, no. 1–2 (January 2, 2014): 64–70.

20. Ana Valens, "Steam's Bestselling, Big-Budget Porn Game Has Got Nothing on These Queer Games," Polygon, April 13, 2021, https://www.polygon.com/22381939 /subverse-steam-kickstarter-review-early-access-porn-sex-adult-games; Alyson Krueger, "Virtual Reality Gets Naughty," *New York Times*, October 28, https://www.nytimes .com/2017/10/28/style/virtual-reality-porn.html.

21. Sabine Harrer, Simon Nielsen, and Patrick Jarnfelt, "Of Mice and Pants: Queering the Conventional Gamer Mouse for Cooperative Play," in *Extended Abstracts of the 2019 CHI Conference on Human Factors in Computing Systems* (Glasgow: ACM, 2019), 1–11.

22. Lynn Comella (chair), "Sex Tech and the Erotic Imaginary: Mediating Intimacies Online and Off," panel presentation to the Society of Cinema and Media Studies Conference, online event, March 2021.

23. Caitlin Donohue, "Center of Sex and Culture Closes—but Dr. Carol Queen Looks to the Future," 48 Hills, January 29, 2019, https://48hills.org/2019/01/center -of-sex-and-culture-closes-but-dr-carol-queen-looks-to-the-future/.

24. Gayle S. Rubin, "Thinking Sex: Notes for a Radical Theory of the Politics of Sexuality," in *Deviations: A Gayle Rubin Reader* (Durham, NC: Duke University Press, 2011), 137–181, 149.

25. Rubin points to the 1978 publication of *The History of Sexuality* as marking this turning point in thinking about the history of sex. Rubin, "Thinking Sex," 149; Michel Foucault, *The History of Sexuality* (New York: Pantheon Books, 1978).

26. Although many histories of sex and sexual devices begin by looking at the dildo and/or the vibrator, I am thinking specifically of the following: Lynn Comella, *Vibrator Nation: How Feminist Sex-Toy Stores Changed the Business of Pleasure* (Durham, NC: Duke University Press, 2017); Hallie Lieberman, *Buzz: A Stimulating History of the Sex Toy* (New York: Pegasus Books, 2017); Rachel P. Maines, *The Technology of Orgasm: "Hysteria," the Vibrator, and Women's Sexual Satisfaction* (Baltimore: Johns Hopkins University Press, 1999).

27. This reference to dildo manufacturers at the start of the 1900s draws from the same advertisement mentioned in note 16. This advertisement clearly states that the dildo being sold is for use either by "men who are tired" or "women who want to play the role of a man."

28. See, for example, Devlin, *Turned On*, 23.

29. Comella, *Vibrator Nation*.

30. Maines, *Technology of Orgasm*.

31. Emily Dreyfuss, "Don't Get Your Valentine an Internet-Connected Sex Toy," *WIRED*, February 14, 2019, https://www.wired.com/story/internet-connected-sex-toys -security/.

32. Ovid, *Metamorphoses*, trans. Rolfe Humphries (Bloomington: Indiana University Press, 2018).

33. Giard explains: "For the moment, no tangible proof of the existence of such dolls has been found. No bill from a merchant, no diary, no engraving, no document that attests that the love doll was produced and marketed before the twentieth century. Most likely the idea of these dolls comes from [the pornographic novelist] Saikaku. But the majority of my interlocutors in Japan, if not all, assert the opposite. 'The first dolls for adults were made in Japan four centuries ago,' they say, repeating what they heard on television or in the press." Agnès Giard, *Un désir d'humain: Les "love doll" au Japon* (Paris: Les Belles Lettres, 2016), 29. I quote Giard at length here because her process of searching for historical traces of the love doll in Japan has many parallels to my search for the dames de voyage.

34. E. T. A. Hoffmann, *Tales of Hoffman* (New York: Penguin Classics, 1982); Auguste Villiers de l'Isle-Adam, *L'Ève future* (Paris: Ancienne Maison Monnier, 1886); Fritz Lang, dir., *Metropolis* (Germany: UFA, 1927).

35. Minsoo Kang, "The Mechanical Daughter of René Descartes: The Origin and History of an Intellectual Fable," *Modern Intellectual History* 14, no. 3 (November 2017): 633–660.

36. Anne McClintock, *Imperial Leather: Race, Gender, and Sexuality in the Colonial Contest* (New York: Routledge, 1995), 138.

37. McClintock, *Imperial Leather*, 139.

38. Anthony Ferguson, *The Sex Doll: A History* (Jefferson, NC: McFarland & Company, 2010), 16.

39. Allison de Fren, "The Anatomical Gaze in Tomorrow's Eve," *Science Fiction Studies* 36, no. 2 (2009): 235–265.

40. Anaïs Nin, *Little Birds* (New York: Harcourt Brace Jovanovich, 1979).

41. Comella, *Vibrator Nation*; Lieberman, *Buzz*; Drucker, *Contraception*; Jessica Borge, *Protective Practices: A History of the London Rubber Company and the Condom Business* (Montreal: McGill-Queen's University Press, 2020); Anjali R. Arondekar, *For the Record: On Sexuality and the Colonial Archive in India* (Durham, NC: Duke University Press, 2009); Cynthia Ann Moya, "Artificial Vaginas and Sex Dolls: An Erotological Investigation" (PhD diss., Institute for Advanced Study of Human Sexuality, 2006); Maines, *Technology of Orgasm*.

42. Giard, *Un désir d'humain*; Josef Nguyen, "Robots, Sex Games, and Queer Processes of Embodying Autonomy," presentation at the Society of Cinema and Media Studies Conference, online event, March 2021; Neda Atanasoski and Kalindi Vora, *Surrogate Humanity: Race, Robots, and the Politics of Technological Futures* (Durham, NC: Duke University Press, 2019); Allison de Fren, "Technofetishism and the Uncanny Desires of A.S.F.R. (alt.sex.fetish.robots)," *Science Fiction Studies* 36, no. 3 (2009): 404–440; Smith, *Erotic Doll*.

43. Julie Wosk, *My Fair Ladies: Female Robots, Androids, and Other Artificial Eves* (New Brunswick: Rutgers University Press, 2015); Anne Marie Balsamo, *Technologies of the Gendered Body: Reading Cyborg Women* (Durham, NC: Duke University Press, 1996); Minsoo Kang, "Building the Sex Machine: The Subversive Potential of the Female Robot," *Intertexts* 9, no. 1 (2005): 5–22.

44. David Parisi, *Archaeologies of Touch: Interfacing with Haptics from Electricity to Computing* (Minneapolis: University of Minnesota Press, 2018); Rachel Plotnick, *Power Button: A History of Pleasure, Panic, and the Politics of Pushing* (Cambridge, MA: MIT Press, 2018); Pozo, "Haptic Media."

45. Mar Hicks, *Programmed Inequality: How Britain Discarded Women Technologists and Lost Its Edge in Computing* (Cambridge, MA: MIT Press, 2017); Janet Abbate, *Recoding Gender: Women's Changing Participation in Computing* (Cambridge, MA: MIT Press, 2012); Ruth Oldenziel, *Making Technology Masculine: Men, Women, and Modern Machines in America: 1870–1945* (Amsterdam: Amsterdam University Press, 2004).

46. Ruha Benjamin, *Race after Technology: Abolitionist Tools for the New Jim Code* (Medford, MA: Polity, 2019); Nettrice R. Gaskins, "Techno-Vernacular Creativity and Innovation across the African Diaspora and Global South," in *Captivating Technology: Race, Carceral Technoscience, and Liberatory Imagination in Everyday Life*, ed. Ruha Benjamin (Durham, NC: Duke University Press, 2019), 253–274; Safiya Umoja Noble, *Algorithms of Oppression: How Search Engines Reinforce Racism* (New York: New York University Press, 2018); Sasha Costanza-Chock, *Design Justice: Community-Led Practices to Build the Worlds We Need* (Cambridge, MA: MIT Press, 2020).

47. Jacqueline Wernimont, *Numbered Lives: Life and Death in Quantum Media* (Cambridge, MA: MIT Press, 2018); Caetlin Benson-Allott, *Remote Control* (New York: Bloomsbury Academic, 2015); Carly Kocurek, *Coin-Operated Americans: Rebooting Boyhood at the Video Game Arcade* (Minneapolis: University of Minnesota Press, 2015).

48. Carlin Wing, "Episodes in the Life of Bounce: Playing with a Rubber Ball," *Cabinet*, no. 56 (Winter 2014–2015), https://www.cabinetmagazine.org/issues/56/wing.php.

49. Sowande' M. Mustakeem, *Slavery at Sea: Terror, Sex, and Sickness in the Middle Passage* (Urbana: University of Illinois Press, 2016); Melody Jue, *Wild Blue Media: Thinking through Seawater* (Durham, NC: Duke University Press, 2020).

50. Nicole Starosielski, *The Undersea Network: Sign, Storage, Transmission* (Durham, NC: Duke University Press, 2015); Alenda Y. Chang, *Playing Nature: Ecology in Video Games* (Minneapolis: University of Minnesota Press, 2019).

51. Jacob Gaboury, "A Queer History of Computing," Rhizome, February 19, 2013, https://rhizome.org/editorial/2013/feb/19/queer-computing-1/.

52. Alexis Pauline Gumbs, *Undrowned: Black Feminist Lessons from Marine Mammals* (Chico, CA: AK Press, 2020); Alexis Lothian, *Old Futures: Speculative Fiction and Queer Possibility* (New York: New York University Press, 2018).

53. Laine Nooney, "A Pedestal, a Table, a Love Letter: Archaeologies of Gender in Videogame History," *Game Studies* 13, no. 2 (2013); Aubrey Anable, *Playing with Feelings* (Minneapolis: University of Minnesota Press, 2018); Whitney Pow, "Outside the Folder, the Box, the Archive: Moving toward a Reparative Video Game History," *ROMchip: A Journal of Game Histories* 1, no. 1 (July 2019), https://romchip.org/index.php/romchip-journal/article/view/76.

54. Nooney, "A Pedestal"; Anable, *Playing with Feelings*, 34.

55. Teddy Pozo, Bonnie Ruberg, and Chris Goetz, "In Practice: Queerness and Games," *Camera Obscura* 32, no. 2 (2017): 153–163.

56. I recognize that there are indeed existing items that we would term *sex robots* that have been developed both in research contexts and for the consumer market. The point that I am making here is that the discourse around sex robots today far exceeds the actual reality of such robots, with many items that are described as extremely "realistic" in fact looking and sounding anything but real.

57. For example, see the 2017 documentary *The Sex Robots Are Coming*, dir. Nick Sweeney.

58. Michael Moran, "Bizarre, 'Hand-Holding' Robot Sounds, Smells and Even Sweats like a Real Woman," *Daily Star*, November 5, 2020, https://www.dailystar.co.uk/tech/news/bizarre-hand-holding-robot-sounds-22962352.

59. Christian Wagner, "Sexbots: The Ethical Ramifications of Social Robotics' Dark Side," *AI Matters Newsletter* 3, no. 2 (2018): 52–58; Laura Bates, "The Trouble with Sex Robots," *New York Times*, July 17, 2017, https://www.nytimes.com/2017/07/17/opinion/sex-robots-consent.html; Blake Foden, "More 'Abhorrent' Child Sex Dolls Imported amid Startling Warning," *Canberra Times*, March 10, 2019, https://www.canberratimes.com.au/story/5993186/more-abhorrent-child-sex-dolls-imported-amid-startling-warning/; Kathleen Richardson, "The Asymmetrical 'Relationship': Parallels between Prostitution and the Development of Sex Robots," *ACM SIGCAS Computers and Society Newsletter* 45, no. 3 (2015): 290–293; Bryan Menegus, "Sex Robots May Literally Fuck Us to Death," *Gizmodo*, December 19, 2016, https://gizmodo.com/sex-robots-may-literally-fuck-us-to-death-1790276123.

60. Ezio Di Nucci, "Robot Sex and the Rights of the Disabled," in *Robot Sex: Social and Ethical Implications*, ed. John Danaher and Neil McArthur (Cambridge, MA: MIT Press, 2018): 73–88; Eduard Fosch-Villaronga and Adam Poulsen, "Sex Robots in Care: Setting the Stage for a Discussion on the Potential Use of Sexual Robot Technologies for Persons with Disabilities," in *Companion of the 2021 ACM/IEEE International Conference on Human-Robot Interaction* (New York: ACM, 2021), 1–9; Wilhelm E. J. Klein and Vivian Wenli Lin, "'Sex Robots' Revisited: A Reply to the Campaign against Sex Robots," *SIGCAS Computers and Society Newsletter* 47, no. 4. (2018): 107–121.

61. Oliver Korn, Gerald Bieber, and Christian Fron, "Perspectives on Social Robots: From the Historic Background to an Expert's View on Future Developments," in *Proceedings of the 11th PErvasive Technologies Related to Assistive Environments Conference* (New York: ACM, 2018): 186–193.

62. I am referring here to the International Congress on Love and Sex with Robots (controversies around which are too tortuous to fully elaborate on here), now being hosted for its sixth year in August 2021. See https://www.lovewithrobots.com/.

63. Hillel Schwartz, *The Culture of the Copy: Striking Likenesses, Unreasonable Facsimiles* (New York: Zone Books, 2014), 129.

64. Safiya Umoja Noble, "Your Robot Is Not Neutral," in *Your Computer Is on Fire*, ed. Thomas S. Mullaney, Benjamin Peters, Mar Hicks, and Kavita Philip (Cambridge, MA: MIT Press, 2021), 199–212.

65. Nguyen, "Robots, Sex Games, and Queer Processes."

66. Sebastien Deterding, "#boycottACE and Institutional Corruption," Gamification Research Network, October 31, 2018, http://gamification-research.org/2018/10/boycottace-and-institutional-corruption/; Marina Adshade, "We Need Academic Conferences about Robots, Love, and Sex," *Slate*, December 13, 2018, https://slate.com/technology/2018/12/love-sex-robots-conference-bannon-academic-research.html.

67. I have chosen not to cite examples directly here, since many of the works in question come from fields in which research "impact" is judged quantitatively by the number of publication citations (regardless of the context of the citation), and these impact scores themselves function as cultural capital, lending legitimacy to such publications.

68. Ari Schlesinger, "A Feminist Programming Language?," FemTechNet, July 14, 2014, https://femtechnet.org/2014/07/a-feminist-programming-language/.

Chapter 1

1. David Levy, *Love and Sex with Robots: The Evolution of Human-Robot Relations* (New York: HarperCollins, 2007).

2. Levy, *Love and Sex with Robots*, 22.

3. Rachel P. Maines, *The Technology of Orgasm: "Hysteria," the Vibrator, and Women's Sexual Satisfaction* (Baltimore: Johns Hopkins University Press, 1999); Levy, *Love and Sex with Robots*, 221.

4. Levy, *Love and Sex with Robots*, 232, 236.

5. Levy, 237.

6. Levy, 237.

7. Levy, 237.

8. Levy, 181.

9. For a recent example of a text that reprints this image and presents it as an actual photograph of the dames de voyage, see Rebecca Clark, "Gag Reflexes: Sex Doll Slapstick and Fran Ross's Oreo," *Post45*, January 22, 2020, https://post45.org/2020/01/gag-reflexes-sex-doll-slapstick-and-fran-rosss-oreo/.

10. Anthony Ferguson, *The Sex Doll: A History* (Jefferson, NC: McFarland & Company, 2010).

11. Ferguson, *Sex Doll*, 1.

12. Ferguson, 1.

13. Ferguson, 16.

14. Hallie Lieberman, *Buzz: A Stimulating History of the Sex Toy* (New York: Pegasus Books, 2017).

15. Lieberman, *Buzz*, 19.

16. Lieberman, 119.

17. Lieberman, 119.

18. Jia Tolentino, "The Rage of the Incels," *New Yorker*, May 15, 2018, https://www
.newyorker.com/culture/cultural-comment/the-rage-of-the-incels.

19. Julien Arbois, *Dans le lit de nos ancêtres: Sexualité, moeurs, et vie intime d'autrefois* (Bernay, France: City Editions, 2016).

20. Kate Lister, *A Curious History of Sex* (London: Unbound, 2021).

21. Arbois, *Dans de lit de nos ancêtres*, 212–213; my original French to English translation.

22. For more on the colonial and racialized implications of these narratives, see chapter 5.

23. Priscille Lamure, "Petite histoire de la poupée érotique," *Savoirs d'Histoire* (blog), October 26, 2017, https://savoirsdhistoire.wordpress.com/2017/10/26/petite-histoire
-de-la-poupee-erotique/comment-page-1/.

24. Lamure, "Petite histoire de la poupée érotique"; my original French to English translation.

25. Julie Bech, "A (Straight, Male) History of Sex Dolls," *Atlantic*, August 4, 2014, https://www.theatlantic.com/health/archive/2014/08/a-straight-male-history-of
-dolls/375623/.

26. Bech, "History of Sex Dolls."

27. Bech.

28. "Nazi Sex Dolls: Hitler's Secret Plan to Manufacture Sex Dolls," *Penthouse*, May 2016.

29. "Nazi Sex Dolls."

30. John Danaher and Neil McArthur, eds., *Robot Sex: Social and Ethical Implications* (Cambridge, MA: MIT Press, 2017), 1.

31. John Danaher, "Should We Be Thinking about Robot Sex?," in *Robot Sex: Social and Ethical Implications*, ed. John Danaher and Neil McArthur (Cambridge, MA: MIT Press, 2017), 3–14.

32. Danaher, "Should We Be Thinking about Robot Sex?," 3.

33. Ruth Oldenziel, *Making Technology Masculine: Men, Women, and Modern Machines in America: 1870–1945* (Amsterdam: Amsterdam University Press, 2004).

34. Levy, *Love and Sex with Robots*, 237; Lieberman, *Buzz*, 119.

35. Ferguson, *Sex Doll*, 16; Bech, "History of Sex Dolls."

36. Arbois, *Dans de lit de nos ancêtres*, 212–213; Lamure, "Petite histoire de la poupée érotique."

37. Levy, *Love and Sex with Robots*, 237.

38. Lieberman, *Buzz*, 19.

39. Ferguson, *Sex Doll*, 16. As for the statement that France and Spain were at the height of their respective naval empires in the seventeenth century, numerous historical accounts document that the 1600s were actually a time of considerable decline for Spain's naval forces following the Thirty Years' War. For France, the seventeenth century appears to have been a period of expansion in the country's naval history, though what we might call the real historical "height" of the French navy seems to have come under Napoleon in the early 1800s. This clarification is relevant because it underscores the ahistoricity and inaccuracy of many seemingly historical details that surround the tale of the dames de voyage. For an overview of these histories, see, for example, Jeremy Black, *Naval Power: A History of Warfare and the Sea from 1500 Onward* (London: Red Globe Press, 2009). For a historiographic reflection on how narratives about maritime history have themselves been constructed starting in the twentieth century, see A. D. Lambert, "The Construction of Naval History 1815–1914," *International Quarterly Journal of the Society for Nautical Research* 97, no. 1 (2011): 207–224.

40. Arbois, *Dans de lit de nos ancêtres*, 212.

41. Bech, "History of Sex Dolls."

42. "Nazi Sex Dolls"; Lamure, "Petite histoire de la poupée érotique."

43. Levy, *Love and Sex with Robots*, 237; Lieberman, *Buzz*, 119.

44. Ferguson, *Sex Doll*, 1.

45. Arbois, *Dans de lit de nos ancêtres*, 213.

46. Bech, "History of Sex Dolls."

47. Jack Z. Bratich and Heidi M. Brush, "Fabricating Activism: Craft-Work, Popular Culture, and Gender," *Utopian Studies* 22, no. 2 (2011): 238.

48. See, for example, Chad M. Mosher, Heidi M. Levitt, and Eric Manley, "Layers of Leather: The Identity Formation of Leathermen as a Process of Transforming Meanings of Masculinity," *Journal of Homosexuality* 51, no. 3 (2006): 93–123.

49. For the gender dynamics of maker culture, see Andrea Marshall and Jennifer Rode, "Deconstructing Sociotechnical Identity in Maker Cultures," in *Proceedings of the 4th Conference on Gender & IT—GenderIT '18* (New York: ACM Press, 2018), 91–100. For the gender dynamics of hacker culture, see Allison Adam, "Hacking into Hacking: Gender and the Hacker Phenomenon," in *Gender, Ethics and Information Technology* (London: Palgrave Macmillan, 2005), 128–146.

50. Bech, "History of Sex Dolls"; Lieberman, *Buzz*; Clark, "Gag Reflexes."

51. Levy, *Love and Sex with Robots*, 301.

52. Ferguson, *Sex Doll*, 16.

53. Another example of a publication that repeats the tale of the dames de voyage during this period is Meredith Gwynne Fair Worthen, ed., *Sexual Deviance and Society: A Sociological Examination* (New York: Routledge, 2016), 262.

54. There are some exceptions, including Lieberman's 2017 *Buzz*, which draws directly from Levy's *Love and Sex with Robots* rather than Ferguson's *Sex Doll*. Lieberman, *Buzz*, 119.

55. Danaher, "Should We Be Thinking about Robot Sex?," 12.

56. Many pages after Ferguson discusses the dames de voyage (primarily on page 16), he describes the so-called Dutch wives supposedly used by the Japanese, which he writes "[originated] in the seventeenth century, when merchant ships would carry leather dolls around for the comfort of the crew" (27). Bech, in her article, pulls these two elements of Ferguson's text together, talking about both the dames de voyage and the Dutch wives in one paragraph. It appears that Danaher then further splices these elements together, creating the "fact" that the dames de voyage were used by seventeenth-century Dutch sailors. Bech also mistakenly—though perhaps intentionally humorously—refers to the dames de voyage as "masturbation puppets," likely a mistranslation of the French word *poupée* as contained in the phrase *poupée erotique* (sex doll)—a signal that Bech is pulling, at least in part, from a French-language source.

57. Michel Foucault, *The History of Sexuality, Volume 1: An Introduction* (New York: Vintage Books, 1990).

58. Sara Ahmed, *Living a Feminist Life* (Durham, NC: Duke University Press, 2017), 15; Amanda Phillips, "Negg(at)Ing the Game Studies Subject," *Feminist Media Histories* 6, no. 1 (January 1, 2020): 16.

59. An entire section in *Robot Sex* is dedicated to the theme of defending robot sex, followed shortly thereafter by another about the possibility of robot love.

60. Among a handful of tells in the *Penthouse* article, the author refers to the sailors' dolls as both dames de voyage and damas de viaje, as Ferguson does.

61. Ferguson, *Sex Doll*, 16, 214. Ferguson gives the title of this magazine as *Avantoure: Anthology of Temptation* in the body of the text on page 16, but lists it as *avantoure*, specifying that the article appeared in the July–August 2006 issue, in his bibliography on page 214. Following Ferguson, and given the inaccessibility of the published article, I cite it here as Amy Wolf, "Dames de Voyage," *avantoure* (July–August 2006): page numbers unknown.

62. Ferguson, *Sex Doll*, 16.

63. Richard Macmanus, "Avantoure: A Magazine for the Web Age," ReadWrite, December 4, 2006, https://readwrite.com/2006/12/04/avantoure_web_magazine/.

64. Personal email correspondence with the author, November 2017.

65. Amy Wolf, "On Water," early draft for "Dames de Voyage" shared with permission by the author (2006).

66. Ferguson, *Sex Doll*, 16–20; Levy, *Love and Sex with Robots*, 177–181, 237–239; Iwan Bloch, *Das Sexualleben unserer Zeit in seinen Beziehungen zur modernen Kultur* (Berlin: Louis Marcus Veriagsbuchhandlung, 1907); Iwan Bloch, *The Sexual Life of Our Time in Its Relations to Modern Civilization*, trans. M. Eden Paul (London: Rebman Limited, 1909); René Schwaeblé, *Les Détraquées de Paris: Étude Documentaire* (Paris: Bibliothèque du Fin de siècle, 1904); Henry N. Cary, *Erotic Contrivances: Appliances Attached to, or Used in Place of, the Sexual Organs* (Chicago: printed privately, 1922). For others seeking out Bloch's text, note that both Ferguson and Levy give slightly incorrect citations. Levy states that the German text dates from 1909; this is in fact the date of the English translation. Ferguson lists the date as 1908, which seems to split the difference between the 1907 German-language and 1909 English-language publications. I offer this clarification because Bloch published a number of books released in a number of editions during this period, so it's particularly useful to have precise dates. For those seeking out Schwaeblé's text, note that Levy lists it as having been published in 1905, when it was in fact published in 1904 and then published in a second edition in 1910. Also, the English-language translations of Schwaeblé's text that Levy provides, which he states come from a translation by John Snugden (Levy, *Love and Sex with Robots*, 179), seem in fact to be amateur translations that have not been published elsewhere.

67. For example, in Ferguson's discussion of sex robots, where he engages with Levy's book at length. Ferguson, *Sex Doll*, 190–193.

68. Levy, *Love and Sex with Robots*, 180.

69. Cynthia Ann Moya, "Artificial Vaginas and Sex Dolls: An Erotological Investigation" (PhD diss., Institute for Advanced Study of Human Sexuality, 2006).

70. David Levy, "Intimate Relationships with Artificial Partners" (PhD diss., Maastricht University, 2007).

71. Levy includes a thank-you to Moya in the acknowledgments section of his book (Levy, *Love and Sex with Robots*, vii) and the preface to his dissertation (Levy, "Intimate Relationships with Artificial Partners," v) for sharing "helpful advice on early 20th century sex artefacts." In our personal correspondence (August 2020), Moya confirmed that she shared her full dissertation project with Levy while he was in the process of writing his own.

72. For instance, Levy's quote from Bloch's *The Sexual Life of Our Time* (Levy, *Love and Sex with Robots*, 177–178) is, in fact, an abbreviated passage from one found in Moya's dissertation (Moya, "Artificial Vaginas and Sex Dolls," 43). Similarly, the extended passage from Cary's *Erotic Contrivances* that Levy quotes (Levy, *Love and Sex with Robots*, 238–239) is itself an excerpt from a longer quotation that Moya provides (Moya, "Artificial Vaginas and Sex Dolls," 62–64). In her own discussion of Cary's text, Moya explains that she found the text in her archival research at the Kinsey Institute Library only by luck. "Possibly less than ten copies" were ever made, she writes, and an error in cataloging meant that even the copy owned by the Kinsey Institute did not come up through standard searches (Moya, "Artificial Vaginas and Sex Dolls," 53). This makes the chances that Levy too was able to access the text extremely limited. Indeed, in the preface of his dissertation, Levy thanks Moya for "kindly provided abstracts" from Cary's text (Levy, "Intimate Relationships with Artificial Partners," v), a note that curiously disappears from the acknowledgments page when Levy's project is revised and published as a book (Levy, *Love and Sex with Robots*, vii).

73. Moya, "Artificial Vaginas and Sex Dolls," 178; Levy, *Love and Sex with Robots*, 237.

74. There is admittedly a complication in this genealogy, since Amy Wolf's article also describes the dames de voyage as sailors' dolls, and was published in 2006, one year before Levy's book. One possible explanation is that Wolf, who mentions no sources in her article, read or heard a version of Levy's work in progress. Another possibility is that both Wolf and Levy were drawing in part from similar versions of the tale circulating through informal early internet spaces like forums and listservs.

75. Levy, *Love and Sex with Robots*, 181.

76. Levy, "Intimate Relationships with Artificial Partners," 186.

77. Levy, *Love and Sex with Robots*, vii; Levy, "Intimate Relationships with Artificial Partners," v.

78. Moya, "Artificial Vaginas and Sex Dolls," 44.

79. Moya, 43.

80. Moya, 44.

81. *Ergänzungswerk zur Sittengeschichte des Lasters: illustrationskommentar für studienbibliotheken und wissen-schaftler/mediziner und juristen zu den textabhandlungen des hauptbandes* (Vienna: Verlag für Kulturforschung, 1927); Leo Schidrowitz, *Sittengeschichte des Lasters die Kulturepochen und ihre Leidenschaften* (Vienna: Verlag für Kulturforschung, 1927); Moya, "Artificial Vaginas and Sex Dolls," 44.

Chapter 2

1. One text from an intermediary time period that I admittedly found late in my research is Hillel Schwartz's book *The Culture of the Copy*, which was originally published in 1996. Schwartz does describe early sex dolls as having been used by sailors and recounts a bawdy anecdote about a sea captain who caught his first mate having sex with a doll. However, the types of dolls that Schwartz is describing are commercially produced rubber dolls much closer to the *femmes en caoutchouc* discussed in chapter 4, and his reference to the use of such dolls by sailors (which he draws from a 1900 essay that introduces a volume on the history of the mannequin) is in keeping with the cultural association between sex dolls and sailors discussed in chapter 3. Thus, it's unlikely that the notion that sailors actually made such dolls at sea came directly to later authors, like Amy Wolf in 2006, through Schwartz, since their stories are so dissimilar. Hillel Schwartz, *The Culture of the Copy: Striking Likenesses, Unreasonable Facsimiles* (New York: Zone Books, 2014); Octave Uzanne, "Les femmes dociles: Visite à l'industriel d'Anvers," introduction to *Le Mannequin*, by Léon Riotor (Paris: Bibliothèque artistique et littéraire, 1900).

2. The texts I focus on in this section are from the 1960s and 1970s, but it's worth noting that a few examples from this genre of work do continue to appear up through the 1990s. One is the 1992 book titled *The Encyclopedia of Unusual Sexual Practices* by Brenda Love, which includes one of the few references to the dames de voyage in her entry for fornicatory dolls, seemingly suggesting that the term *dames de voyage* might serve as a synonym for sex dolls more generally. Love does also mention sailors in this entry, stating that "the more expensive types [of sex dolls] were once popular among European sailors and had pubic hair and even a clitoris"—a suspect claim followed not by a source but instead, curiously, by a description of an inflatable sheep with a hole beneath its tail for penetration. Brenda Love,

The Encyclopedia of Unusual Sexual Practices (New York: Barricade Books, 1992), 118–119.

3. Alfred C. Kinsey, Wardell B. Pomeroy, and Clyde E. Martin, *Sexual Behavior in the Human Male* (Bloomington: Indiana University Press, 1998); Alfred C. Kinsey, Wardell B. Pomeroy, Clyde E. Martin, and Paul H. Gebhard, *Sexual Behavior in the Human Female* (Bloomington: Indiana University Press, 1998).

4. Advertising images are from the following: Paul Tabori, *The Humor and Technology of Sex* (New York: Julian Press, 1969), n.p.; Evelyn Rainbird, *The Illustrated Manual of Sexual Aids* (New York: Minotaur Press, 1973), 68.

5. An example of one such text that focuses explicitly on so-called fornicatory dolls and related items is Aaron J. Abelard, *Substitute Lovers* (Hollywood: Barclay House, 1969).

6. R. von Krafft-Ebing, *Psychopathia Sexualis, with Special Reference to Contrary Sexual Instinct: A Medico-Legal Studies*, trans. Charles Gilbert Chaddock (Philadelphia: F. A. Davis Company, 1894); Havelock Ellis, *Studies in the Psychology of Sex* (New York: Random House, 1936).

7. The two books with the same title published in the same year are Gerhard Stoltz, *Sex Gadgets* (Cleveland, Ohio: Classics Library, 1968) and Roger Blake, *Sex Gadgets* (Cleveland, Ohio: Century, 1968). The quote is from Cynthia Ann Moya, "Artificial Vaginas and Sex Dolls: An Erotological Investigation" (PhD diss., Institute for Advanced Study of Human Sexuality, 2006), 12.

8. Amy Dumont and Ashton Dumont, *Sex Devices and How to Use Them* (Los Angeles: Argyle Books, 1970).

9. Dumot and Dumont, *Sex Devices*, 5.

10. Dumot and Dumont, 3–56, 60–64.

11. Tabori, *Humor and Technology of Sex*, 383–384. The reference to oil here suggests that Tabori is drawing from Iwan Bloch's work, as described ahead, or from another author who drew from Bloch.

12. Evelyn Rainbird, *Illustrated Manual of Sexual Aids*. Ironically, despite the higher quality of research, Rainbird is one of the most mysterious authors in this bunch; Moya points out that she is likely a "made-up character," probably a creation of the Penthouse Media Group, though she is far from the only author mentioned here to use a pseudonym. Moya, "Artificial Vaginas and Sex Dolls," 83.

13. Rainbird, *Illustrated Manual of Sexual Aids*, 49.

14. Rainbird, 60.

15. Dumont and Dumont, *Sex Devices*; Roger Blake, *Sex Gadgets*; Tabori, *Humor and Technology of Sex*.

16. Jane Long, *A Housewife's Guide to Auto-Erotic Devices in the Home* (San Diego: Greenleaf Classics, 1972). Rainbird also mentions this book, even including an image of the cover in her book, with the description, "The cover of a typical American paperback, ostensibly giving unusual data but in fact consisting of casually invented fantasy material." Rainbird, *Illustrated Manual of Sexual Aids*, 63.

17. Henry N. Cary, *Erotic Contrivances: Appliances Attached to, or Used in Place of, the Sexual Organs* (Chicago: printed privately, 1922); Albert Ellis and Albert Abarbanel (eds.), *The Encyclopedia of Sexual Behavior: Volume One* (New York: Hawthorn Books, 1961), 151.

18. Heike Bauer, "Disciplining Sex and Subject: Translation, Biography and the Emergence of Sexology in Germany," in *English Literary Sexology* (London: Palgrave Macmillan, 2009), 21–51.

19. Laurie Marhoefer, *Sex and the Weimar Republic: German Homosexual Emancipation and the Rise of the Nazis* (Toronto: University of Toronto Press, 2015).

20. Heike Bauer, *The Hirschfeld Archives: Violence, Death, and Modern Queer Culture* (Philadelphia: Temple University Press, 2017).

21. Leo Schidrowitz, *Sittengeschichte des Lasters die Kulturepochen und ihre Leidenschaften* (Vienna: Verlag für Kulturforschung, 1927); Leo Schidrowitz, *Ergänzungswerk zur Sittengeschichte des Lasters* (Vienna: Verlag für Kulturforschung, 1927).

22. Schidrowitz, *Sittengeschichte des Lasters*, 214. My original translation from the German.

23. Erich Wulffen, *Der Sexualverbrecher* (Berlin: Langenscheidt, 1910).

24. Wulffen, *Der Sexualverbrecher*, 499.

25. Georg Back, *Sexuelle Verirrungen des Menschen und der Natur* (Berlin: Standard-Verlag, 1910), 432–433.

26. Exactly which text of Bloch's these slightly later works were drawing from gets a little messy. Bech is clearly quoting from Bloch's 1907 *The Sexual Life of Our Time*, discussed ahead. Wulffen, by contrast, offers slightly different information, but still credited to Bloch. This is because Wulffen appears to actually be drawing from an earlier discussion of the dames de voyage that appeared in Bloch's 1903 book, *Beiträge zur Aetiologie der Psychopathia Sexualis* [Contributions to the aetiology of psychopathia sexualis]. Iwan Bloch, *Beiträge zur Aetiologie der Psychopathia Sexualis: Zweiter Teil* (Dresden: Verlag von H. R. Dohrn, 1903).

27. Moya, "Artificial Vaginas and Sex Dolls," 37–38.

28. Iwan Bloch, *Das Sexualleben unserer Zeit in seinen Beziehungen zur modernen Kultur* (Berlin: Louis Marcus Veriagsbuchhandlung, 1907). English-language texts sometimes state the date of this work as 1909; that is because 1909 is the publication date for the official English translation of Bloch's book: Iwan Bloch, *The Sexual Life of Our Time in Its Relations to Modern Civilization*, trans. M. Eden Paul (London: Rebman Limited, 1909).

29. In addition to texts that use this passage to talk explicitly about the dames de voyage, we see this same passage from Bloch in other works about sex tech and sex dolls, such as Allison de Fren, "Technofetishism and the Uncanny Desires of A.S.F.R. (alt.sex.fetish.robots)," *Science Fiction Studies* 36, no. 3 (2009): 401–440, 410.

30. Bloch, *Sexual Life of Our Time*, 648–649.

31. Although *The Sexual Life of Our Time* is often the text credited with originating Bloch's discussion of the dames de voyage, Bloch in fact first writes about the dames de voyage four years earlier, in his 1903 book, *Beiträge zur Aetiologie der Psychopathia Sexualis* [Contributions to the aetiology of psychopathia sexualis]—which did not, unlike *The Sexual Life of Our Time*, receive an official English translation. Here, in this earlier text, Bloch positions his discussion of the dames de voyage after a description of people who have sex with statues. He then writes: "How far fornication goes in this area is shown by the fact that today the so-called 'dames de voyage,' i.e., whole female bodies made of rubber, are sold to debauched old men [*roués*]. The genitals are faithfully imitated and even the secretion of the glandulae Bartholini is imitated by a 'pneumatic tube' filled with oil. There are even said to be replicas of full men for women." Bloch, *Beiträge zur Aetiologie der Psychopathia Sexualis*, 301; my original translation. Bloch later revises and expands upon this earlier description in his 1907 text, adding more details about the dolls' workings and citations to related texts while also removing the word *roués*, a disparaging term—in effect, casting the purchase and use of elaborate sex dolls in a more positive light.

32. For examples of works representing the recurring interest in the Digesting Duck from scholars in science and technology studies, see Daniel Cottom, "The Work of Art in the Age of Mechanical Digestion," *Representations* 66 (April 1, 1999): 52–74; Jessica Riskin, "The Defecating Duck, or, the Ambiguous Origins of Artificial Life," *Critical Inquiry* 29, no. 4 (June 2003): 599–633.

33. For example, see Anthony Ferguson's comment that sex dolls sprang from "the germ of male desire" to "create something which was female in appearance, but completely receptive and non-judgmental." Ferguson, *Sex Doll*, 1.

34. Note that this *as* construction is consistent in Bloch's original German text (which uses *als*) as well as the official English translation. Bloch, *Das Sexualleben unserer Zeit*, 710.

35. Admittedly, the fact that Bloch chooses to use the French terms *hommes* and *dames de voyage* in his German-language text adds another layer of complexity, suggesting that the terms may be euphemistic or idiomatic.

36. Bloch, *Sexual Life of Our Time*, 649. Others translate this title as "The Misfits of Paris," but the meaning of *détraqué* is closer to the English *deranged*, communicating a sense of wild (in this case, sexual) behavior. The second *e* in *détraquées* as it appears in Schwaeblé's title implies that the book is specifically about Parisian women engaging in wild behavior.

37. Bloch, *Sexual Life of Our Time*, 649. The word that Bloch uses to describe Madame B.'s story in his original German text is not *romance* (as the English translation suggests) but rather *roman*, a novel. This distinction, while minor, helps make clear that Bloch did recognize the fictional nature of the work, while later authors who have learned about *La Femme endormie* through the English translation of Bloch's text have often interpreted the phrase *erotic romance* to mean an actual, factual account of an extended romantic tryst between a human man and a sex doll. Bloch, *Das Sexualleben unserer Zeit*, 711.

38. Bloch, *Sexual Life of Our Time*, 649.

39. René Schwaeblé, *Les Détraquées de Paris: Étude Documentaire* (Paris: Bibliothèque du Fin de siècle, 1904). A second edition of the book was released in 1910, this time without the illustrations—supposedly so that the volume would be deemed less pornographic and more appropriate for sale: René Schwaeblé, *Les Détraquées de Paris: Étude de moeurs contemporaines* (Paris: H. Daragon, 1910).

40. As blogger Sabine Huet points out in a 2008 post, *Les Détraquées de Paris* could be considered a queer text, part of what Huet refers to as a largely missing history of lesbian erotica. Sabine Huet, "René Schwaeblé, 'Les détraquées de Paris'. Etude de mœurs contemporaines.' Nouvelle Edition, Daragon libraire-éditeur, 1910," *Les Introuvables lesbiens* (blog), December 3, 2008, http://romanslesbiens.canalblog.com /archives/2008/12/03/7575073.html.

41. Victor Leca, *Paris-Fêtard: Guide secret de tous les plaisirs* (Paris: P. de Porter, 1907).

42. Leca, *Paris-Fêtard*, 21.

43. *Le Rire: Journal humoristique*, October 29, 1910.

44. Schwaeblé, *Les Détraquées de Paris: Étude de moeurs contemporaines*, 84–85. Note that page numbers for *Les Détraquées de Paris* come from the 1910 edition.

45. Schwaeblé, *Les Détraquées de Paris*, 33.

46. Schwaeblé, *Les Détraquées de Paris*, 33. Another charming bit of humor in this story is that, the narrator claims, Dr. P. has to take pains to avoid arrest and make

sure that his bespoke sex doll fabrication operation looks like a legitimate business, so he stocks his store with large balloon animals to throw police off the trail.

47. I am thinking here specifically of the anecdote told by Octave Uzanne in his introduction to *Le Mannequin*, where he describes having visited the city of Antwerp fifteen years prior and been taken to a shop that sold hand-crafted life-sized dolls. This story too has a bawdy air to it, like a tall tale of sex tourism. Uzanne, "Les femmes dociles."

48. Bloch, *Sexual Life of Our Time*, 649.

49. Bloch, 649.

50. Madame B. (Alphonse Momas), *La Femme endormie* (Melbourne: J. Renold, 1899).

51. See, for example, the following texts, all authored by Momas: Georges de Lesbos, *Voluptés bizarres* (Amsterdam: no publisher listed, 1893); Erosmane, *Lubricités, récits intimes et véridiques d'ancedotes galantes extraites de la vie privée des célébrités contemporaines* (Brussels: no publisher listed, 1891); Fuckwell, *Petites et grandes filles* (London: no publisher listed, 1907).

52. Jon Stratton, *The Desirable Body: Cultural Fetishism and the Erotics of Consumption* (Manchester: Manchester University Press, 1996), 215; Auguste Villiers de l'Isle-Adam, *L'Ève future* (Paris: Ancienne Maison Monnier, 1886).

53. Paul Booth, "Slash and Porn: Media Subversion, Hyper-Articulation, and Parody," *Continuum* 28, no. 3 (May 4, 2014): 396–409.

54. Madame B., *La Femme endormie*, 1.

55. Madame B., 1.

56. Madame B., 5.

57. Madame B., 3.

58. Madame B., 3.

59. Bloch, *Sexual Life of Our Time*, 649. Interestingly, Bloch does not mention these catalogs in his 1903 book, *Beiträge zur Aetiologie der Psychopathia Sexualis* [Contributions to the aetiology of psychopathia sexualis], in which an earlier version of his discussion of the dames de voyage appears, suggesting that he encountered these catalogs sometime between 1903 and 1907.

60. Bloch, *Sexual Life of Our Time*, 649.

61. Many of the documents surrounding the dames de voyage suggest that the sale of sex dolls was illegal in Paris during this period. See, for example, *Les Détraquées de Paris*, which includes reference to the fact that the seller of sex dolls

in the "Homunculus" story had to avoid drawing the attention of the police. Other evidence includes documentation of legal cases in which individuals who distributed advertisements for sex dolls on the streets of Paris were brought to court under charges related to the promotion of obscenity and indecency, as discussed in chapter 4.

62. A small selection of catalogs for rubber goods and other intimate devices has thankfully been preserved by the Internet Archive. Other examples come from Album 7 in the Milford Haven Collection, currently held by the British Library, as described ahead.

63. Advertisements for these catalogs begin appearing in Parisian newspapers as early as 1891, starting with the rubber manufacturer Maison A. Claverie. Claverie's ads are followed soon after by many other sellers, with names such as Maison Durand, Maison C. Bor, Leigh's, and Office des Inventions Reunies, though it's unclear how many of these companies apart from Claverie actually produced their own items.

64. Manuel Charpy, "Craze and Shame: Rubber Clothing during the Nineteenth Century in Paris, London, and New York City," *Fashion Theory* 16, no. 4 (December 2012): 433.

65. Cary, *Erotic Contrivances*, 48. Cary's reference here to *naval officers* does not appear in Rainbird's text. However, it's consistent with the cultural associations between sex dolls and sailors that started forming in France at the turn of the twentieth century, which I discuss in chapter 3.

66. Rainbird, *Illustrated Manual of Sexual Aids*, 53.

67. Rainbird, 53.

68. Tracking down these documents has been a saga in and of itself, and I detail it here so that others after me will hopefully be able to find them. Rainbird gives the following citation for her reference to these sex toy catalogs: "British Museum Private Case, Album 7, The Milfordhaven Catalogues" (Rainbird, *Illustrated Manual of Sexual Aids*, 51). This turns out to be a citation to what is referred to as the Milford Haven collection, which was previously the private collection of George Mountbatten, second Marquess of Milford Haven. Originally, the collection contained pornography, erotic catalogs, and erotic postcards. When George Mountbatten died in 1938, the collection passed to his son, David Mountbatten. In 1961, David Mountbatten was implicated in the Profumo affair, a major political scandal in which Secretary of the State for War John Profumo (under conservative prime minister Harold Macmillan) was revealed to be having an extramarital affair with a nineteen-year-old model. In the aftermath, David Mountbatten dispersed his father's collection. The curators at the British Library report that, of this initial collection, "a small collection of erotic prospectuses and catalogs for obscene books, pictures and instruments,

dating from 1889 to 1929 was donated to the British Museum in 1963 and these are now at Cup.364.g.48." Email communication from Elias Mazzucco on the Rare Books and Music Reference Team at the British Library, April 16, 2021. These items were stored briefly in the British Museum's infamous Private Case, established in the 1850s to contain "obscene material," the contents of which were later transferred to the British Library in 1973. It is there at the British Library, still at Cup.364.g.48, that these items still remain, where they are overseen by the Rare Books team. Unfortunately, they have been deemed too delicate to produce images of, which is why the images from the catalogs here in this book have been recreated by an artist. Those on the hunt for these materials will find that a second segment of George Mountbatten's original collection—namely, his cache of erotic postcards—has landed at the Victoria and Albert Museum, where it's also labeled as the Milford Haven Collection. As for the actual catalogs of sexual devices in question, they are contained in what is called Album 7. In 2019, bibliographer Patrick J. Kearney published a list of the contents of Album 7. Among these are advertisements and catalogs for erotic books, but some are for clandestine catalogs of erotic novelties and apparatuses. The full citation for the bibliographic list of the contents of Album 7 is as follows: Patrick J. Kearney, *Album 7: A Transcription of an Important Collection of Erotic Ephemera in the British Library* (Santa Rosa, CA: Scissors & Paste Bibliographies, 2019). For more information about the humorous circumstances by which the Milford Haven postcard collection came to be archived by the Victoria and Albert Museum (one begrudging internal memo between curators written in 1982 reads, "I supposed we are committed to accepting this gift"), see Erika Lederman, "French Postcards: History Revealed," *V&A Blog*, November 2, 2015, https://www.vam.ac.uk/blog/caring-for-our -collections/french-postcards-history-revealed?doing.

69. Advertisement (bottom-right corner of page), *La Lanterne*, June 25, 1891, 4; my French to English translation of this advertisement.

70. Advertisement (bottom center), *La Grisette: Revue populaire illustrée*, August 10, 1895, 3; my French to English translation of this advertisement.

71. Maison L. Bador, *Fabrication perfectionée de caoutchouc dilaté et baudruche* (Paris: commercial catalog, 1900), cover.

72. Maison L. Bador, *Fabrication perfectionée*, 1; my original French to English translation.

73. Maison L. Bador, 8; my original French to English translation.

74. Maison L. Bador, 10; my original French to English translation.

75. Jessica Borge, *Protective Practices: A History of the London Rubber Company and the Condom Business* (Montreal: McGill-Queen's University Press, 2020), 17.

76. Agnès Giard, *Un désir d'humain: Les "love doll" au Japon* (Paris: Les Belles Lettres, 2016), 29.

Chapter 3

1. Jean-Yves Mollier, *Le Camelot et la rue: Politique et démocratie au tournant des XIXe et XXe siècles* (Paris: Fayard, 2004), 7.

2. Margaret C. Creighton and Lisa Norling, "Introduction," in *Iron Men, Wooden Women: Gender and Seafaring in the Atlantic World, 1700–1900*, ed. Margaret C. Creighton and Lisa Norling (Baltimore: Johns Hopkins University Press, 1996), vii–xiv, viii.

3. Mollier, *Le Camelot et la rue*, 8.

4. M. le compte E. A. Sallins, "En Des Jours pareils! Récit russe du temps de Pougatcheff," in *Bibliothèque universelle et revue Suisse* (Lausanne, Switzerland: Bureaux de la Bibliothèque universelle, October 1893), 149–167. The story's subtitle, "A Russian story from the time of Pougatcheff," refers to a mid-eighteenth-century pretender to the Russian throne who led an unsuccessful rebellion against Catherine II. It's therefore best understood as a work of loosely historical fiction.

5. Sallins, "En Des Jours pareils!," 154–155.

6. Admittedly, the term does not seem to have been one of the more commonly used colloquial terms for sex workers. *Paris-Fêtard*, the 1907 guide to Parisian sex tourism that I discuss in chapter 2, refers to prostitutes and women performing other kinds of erotic labor using a variety of lingo (e.g., *dames galantes* or "loose women"), but nowhere does it mention *dames de voyage*. Victor Leca, *Paris-Fêtard: Guide secret de tous les plaisirs* (Paris: P. de Porter, 1907).

7. "Le Rire de la semaine," *Le Rire: Journal humoristique*, March 19, 1910, no listed page numbers (pages 1 and 2 following cover).

8. The year 1903 is the publication date for Iwan Bloch's *Beiträge zur Aetiologie der Psychopathia Sexualis: Zweiter Teil*, in which he first discusses the dames de voyage. Iwan Bloch, *Beiträge zur Aetiologie der Psychopathia Sexualis: Zweiter Teil* (Dresden: Verlag von H. R. Dohrn, 1903), 301.

9. Henry N. Cary, *Erotic Contrivances: Appliances Attached to, or Used in Place of, the Sexual Organs* (Chicago: printed privately, 1922), 48; Evelyn Rainbird, *The Illustrated Manual of Sexual Aids* (New York: Minotaur Press, 1973), 49.

10. "Le Rire de la semaine." Although this story is presented in a humorous light, it appears to be true as it was reported on simultaneously by multiple newspapers. For example, see the write-up of the court case in the Chronique (Chronical) section of *L'Éclat de rire: Journal humoristique*. G. de Saint-Loup, "Chronique," *L'Éclat de rire: Journal humoristique* 134 (1910): 14.

11. "Le Rire de la semaine"; my original French to English translation.

12. "Le Rire de la semaine"; my original French to English translation.

13. "Le Rire de la semaine"; my original French to English translation.

14. René Crevel, *Êtes-Vous Fous?* (Paris: Librairie Gallimard, 1929).

15. Crevel, *Êtes-Vous Fous?*, 178

16. Roger Peyrefitte, *Des Français* (Paris: Flammarion, 1973).

17. Peyrefitte, *Des Français*, 101; my original French to English translation.

18. Vladimir Volkoff, *The Turn-Around*, trans. Alan Sheridan (London: Bodley Head, 1981).

19. Volkoff, *Turn-Around*, 130.

20. Madame B. (Alphonse Momas), *La Femme endormie* (Melbourne: J. Renold, 1899).

21. Georges Eclar, "La Femme du capitaine," *La Grisette: Revue populaire illustrée*, October 10, 1895, 83–85.

22. Eclar, "La Femme du capitaine," 84.

23. Eclar, 84–85.

24. Eclar, 85.

25. Eclar, 84.

26. Clément Voutel, "La Poupée," *La Vie Parisienne*, October 15, 1921, 883, 886–887.

27. As mentioned in the introduction, Olympia appears in Hoffman's 1816 story "Der Sandman" ("The Sandman"), which is about a young man who falls in love with a beautiful automaton, forsaking his flesh-and-blood fiancée and ultimately meeting an untimely demise in his quest to attain her. See E. T. A. Hoffmann, *Tales of Hoffman* (New York: Penguin Classics, 1982).

28. Voutel, "La Poupée," 886; my original French to English translation.

29. Vald'Es, "Portraits en pied," *La Vie Parisienne*, October 15, 1921, 888.

30. For more on these advertisements, see chapter 2. For more on the sale of these items, see chapter 4.

31. Cary, *Erotic Contrivances*, 48; as quoted in Cynthia Ann Moya, "Artificial Vaginas and Sex Dolls: An Erotological Investigation" (PhD diss., Institute for Advanced Study of Human Sexuality, 2006), 62.

32. For more on the various roles that women played in the seafaring world and on ships, see Margaret S. Creighton and Lisa Norling, eds., *Iron Men, Wooden Women:*

Gender and Seafaring in the Atlantic World, 1700–1900 (Baltimore: Johns Hopkins University Press, 1996).

33. Haskell Springer, "The Captain's Wife at Sea," in *Iron Men, Wooden Women: Gender and Seafaring in the Atlantic World, 1700–1900*, ed. Margaret S. Creighton and Lisa Norling (Baltimore: Johns Hopkins University Press, 1996), 92–117.

34. See, for example, Norling and Creighton, *Iron Men, Wooden Women*; Lisa Norling, *Captain Ahab Had a Wife: New England Women and the Whalefishery, 1720–1870* (Chapel Hill: University of North Carolina Press, 2000); David Cordingly, *Seafaring Women: Adventures of Pirate Queens, Female Stowaways, and Sailors' Wives* (New York: Random House, 2007); Suzanne J. Stark, *Female Tars: Women aboard Ship in the Age of Sail* (Annapolis, MD: Naval Institute Press, 2017).

35. For more recent historical writing on women pirates, see Laura Sook Duncombe, *Pirate Women: The Princesses, Prostitutes, and Privateers Who Ruled the Seven Seas* (Chicago: Chicago Review Press, 2017).

36. See, for example: Henry Trotter, "Soliciting Sailors: The Temporal Dynamics of Dockside Prostitution in Durban and Cape Town," *Journal of Southern African Studies* 35, no. 3 (September 2009): 699–713.

37. Sowande' M. Mustakeem, *Slavery at Sea: Terror, Sex, and Sickness in the Middle Passage* (Urbana: University of Illinois Press, 2016), 83–84.

38. See, for example, David Cordingly, "Men without Women," in *Seafaring Women*, 138–153.

39. Cordingly, "Men without Women," 142.

40. Cordingly, 145.

41. See, for example, Daniel Hannah, "Queer Hospitality in Herman Melville's 'Benito Cereno,'" *Studies in American Fiction* 37, no. 2 (2010): 181–201; Hiram Pérez, "The Queer Afterlife of Billy Budd," in *A Taste for Brown Bodies* (New York: New York University Press), 25–48.

42. My thanks to Gina Bardi, reference librarian at the San Francisco Maritime National Historical Park Research Center for confirming my understanding of the particular challenges posed by these documents and the unlikelihood of finding references to sex dolls in such materials. Personal communication with the author, November 9, 2017.

43. Gina Bardi, personal communication with the author, November 9, 2017.

44. One other route I explored at length, but which ultimately proved to be a dead end, was scouring various archives for the Spanish phrase *damas de viaje*, which a

handful of contemporary authors mention alongside dames de voyage. This turns up nothing of relevance; my guess is that the inclusion of damas de viaje into discussions of early sex dolls is a twenty-first-century invention. A heads-up for others after me interested in pursuing a similar line of investigation: the World Newspaper Archive initially seems to turn up numerous references to damas de viaje, but closer inspection reveals these are actually bits of text from either high-society announcements about women going on trips or advertisements for women's luggage.

45. Peter Kasin, "The Monthly Chantey Sing at San Francisco Maritime National Historical Park: An Introduction" (no publication info; shared via personal communication with author, November 11, 2017). The info sheet also includes a note explaining that there are various forms of the word *chantey*, which can apparently be spelled *chantey, chanty, shanty, shantie,* or *shantey*.

46. Kasin, "Monthly Chantey Sing."

47. These include Peter Kasin, park ranger in the Interpretation Division and sea chantey expert at the San Francisco Maritime National Historical Park; Amy Parsons, associate professor in the Department of Culture and Communication at California State University Maritime Academy; and Gibb Schreffler, assistant professor in the Department of Music at Pomona College.

48. Individual correspondence with the author, November 13, 2017.

49. My thanks to Amy Parsons for suggesting this line of archival inquiry and for pointing me to the American Periodicals Series database, from which the primary sources related to this topic were drawn.

50. The Seaman's Friend, "The Salvation of Seamen Difficult," *The Christian Herald and Seaman's Magazine*, March 7, 1824, 153.

51. G. McPherson Hunter, "The Sailor and City Problems: Where the Real Peril of the Seaman Begins," *New York Observer and Chronicle*, May 3, 1906, 562.

52. Hunter, "The Sailor and City Problems," 562.

53. Hunter, 562.

54. Individual correspondence with the author, November 16, 2017.

55. Cordingly writes: "We find the conventional belief that most ships' figureheads depicted women to be far from the case. The predominance of female figureheads was a nineteenth-century phenomenon, and it is only because so many more of these have survived than the earlier lions, dragons, and warriors that we assume that women were more popular on the bows of ships" (Cordingly, *Seafaring Women*, xv).

56. Creighton and Norling, "Introduction," vii, x.

57. Information about this figurehead comes from the catalog entry for the item as part of the collection at the National Maritime Museum. See https://collections.rmg.co.uk/collections/objects/18784.html.

58. Michael P. Dyer, "Scrimshaw," in *Encyclopedia of Marine Mammals*, 3rd ed., ed. Bernd Würsig, J. G. M. Thewissen, and Kit M. Kovacs (Burlington, MA: Elsevier, 2018), 841–845.

59. Some examples include the Mystic Seaport Museum, the Nantucket Whaling Museum, and the Hull Maritime Museum.

60. Janet West and Arthur G. Credland, *Scrimshaw: The Art of the Whaler* (Hull, UK: Hull City Museums & Art Galleries, 1995).

61. "'Whale Bone Porn': Ann Pimental Outraged at Vancouver Maritime Museum's Scrimshaw Exhibit," *HuffPost*, March 25, 2013, https://www.huffpost.com/entry/whale-bone-porn-ann-pimentel_n_2950987.

62. Information about items on display via Gilda Salomone, "There's More to the Vancouver Maritime Museum's Exhibit than 'Whale Bone Porn,'" Radio Canada International, March 25, 2013, https://www.rcinet.ca/en/2013/03/25/theres-more-to-vancouver-maritime-museums-exhibit-than-whale-bone-porn/; Jason Smythe, "I Know It When I See It: Scrimshaw and Whale Bone Porn," *Satellite Gallery* (blog), April 27, 2013, https://satellitegallery.wordpress.com/2013/04/27/i-know-it-when-i-see-it-scrimshaw-and-whale-bone-porn-2/.

63. Tristin Hopper, "'It's All Fake: Vancouver Exhibit's 'Whale Bone Porn' Is Not 19th-Century Scrimshaw, Former Museum Director Says," *National Post*, March 30, 2013, https://nationalpost.com/news/canada/its-all-fake-vancouver-exhibits-whale-bone-porn-is-not-19th-century-scrimshaw-former-museum-director-says.

64. Hopper, "'It's All Fake.'"

Chapter 4

1. For the British context, see, for example, Jessica Borge, *Protective Practices: A History of the London Rubber Company and the Condom Business* (Montreal: McGill-Queen's University, 2020). For the German context, see, for example, Götz Aly and Michael Sontheimer, *Fromms: How Julius Fromm's Condom Empire Fell to the Nazis*, trans. Shelley Laura Frisch (New York: Other Press, 2009); Donna J. Drucker, *Contraception: A Concise History* (Cambridge, MA: MIT Press, 2020), 40–41. For the American context, see, for example, Drucker, *Contraception*, 22.

2. As I will discuss in greater detail when I address issues of race surrounding the tale of the dames de voyage, these rubber items were sometimes manufactured in a dark

brown-black color, as in the example of the *ventre de femme* mentioned in chapter 2. At other times, they were produced in white or pink.

3. Borge, *Protective Practices*, 16.

4. Drucker, *Contraception*, 17.

5. Borge, *Protective Practices*, 16. As Borge helpfully clarifies, rubber vulcanization has been described as a "first revolution" in the manufacture of such items, followed later by another "revolution" in the development of latex rubber, which was thinner, cheaper, stronger, and safer to produce than simple galvanized rubber and which became nearly ubiquitous as the go-to material for condoms by the 1930s. Borge, *Protective Practices*, 17, 19. Drucker, *Contraception*, 17.

6. Carlin Wing, "Episodes in the Life of Bounce: Playing with a Rubber Ball," *Cabinet*, no. 56 (Winter 2014–2015), https://www.cabinetmagazine.org/issues/56/wing.php.

7. Varnout et Galante, *Catalogue des appareils et instruments de médicine et chirurgie en caoutchouc vulcanisé* (Paris: Imprimerie Administrative de Paul Dupont, 1851).

8. Varnout et Galante, *Catalogue des appareils*, title page.

9. Varnout et Galante, 35–36, 15–16, 13–14, 46.

10. Maison L. Bador, *Fabrication perfectionnée de caoutchouc dilaté et baudruche* (Paris: commercial catalog, 1900).

11. Maison L. Bador, *Fabrication perfectionnée*, 3.

12. Maison L. Bador, 16, 12.

13. Advertisement for Maison A. Claverie, *La Lanterne*, June 25, 1891, 4 (lower right-hand corner).

14. Advertisement for Maison A. Claverie, *La Lanterne*, September 20, 1891, 4 (lower middle).

15. Information about the contemporary Claverie store, as well as text and images presenting the company's narrative about its own history, are available at https://mademoiselleclaverie.com/ (accessed April 29, 2021).

16. See https://mademoiselleclaverie.com/histoire/.

17. René Schwaeblé, "Homunculus," in *Les Détraquées de Paris* (Paris: Bibliothèque du Fin de siècle, 1904).

18. Julian Jackson, *Living in Arcadia: Homosexuality, Politics, and Morality in France from the Liberation to AIDS* (Chicago: University of Chicago Press, 2009), 22.

19. Raisa Adah Rexer, *The Fallen Veil: A Literary and Cultural History of the Photographic Nude in Nineteenth-Century France* (Philadelphia: University of Pennsylvania Press, 2021), 4.

20. Rexer, *Fallen Veil*, 23. This reference to paper dolls suggests small, flat, pornographic cutouts, not sex dolls.

21. Rexer, 123.

22. Rexer talks about this phenomenon occurring in the 1870s, with street vendors of nude images "arrested all over Paris" (Rexer, 150). Yet in 1900, four hundred *camelots* (street hawkers) were arrested for carrying obscene postcards. Jean-Yves Mollier, *Le Camelot et la rue: Politique et démocratie au tournant des XIXe et XXe siècles* (Paris: Fayard, 2004), 261.

23. Feona Attwood, "Fashion and Passion: Marketing Sex to Women," *Sexualities* 8, no. 4 (October 2005): 392–406.

24. Carina Hsieh and Natasha Burton, "A Brief History of the Rabbit," *Cosmopolitan*, June 30, 2020, https://www.cosmopolitan.com/sex-love/advice/a4805/history-of-the-rabbit/.

25. As Mark McLelland explains in discussing censorship of pornographic media materials in Japan, "the number of obscenity cases brought before the courts since the 1970s has been relatively small," which McLelland attributes to "a range of self-regulatory mechanisms . . . that advise members on permissible limits." This suggests that the design of the rabbit vibrator may have emerged through similar self-regulatory mechanisms. Mark McLelland, "Sex, Censorship and Media Regulation in Japan: A Historical Overview," in *Routledge Handbook of Sexuality Studies in East Asia*, ed. Mark McLelland and Vera Mackie (London: Routledge, 2014), 409.

26. For an influential example, see David Levy, *Love and Sex with Robots: The Evolution of Human-Robot Relations* (New York: Harper Collins, 2007), 249.

27. The advertisement for a demicorps can be found in Album 7 of the Milford Haven Collection at the British Library. While only a selection of the catalog in which the advertisement appears has been preserved in the album, the seller seems to have left hand-written notes on the preserved pages, indicating that the items are being sold by a company called the Office des Inventions Réunies at Rue Truffaut in Paris.

28. "Les Femmes en caoutchouc," *L'Indépendent de Mascara*, August 27, 1885, 3; Georges Eclar, "La Femme du capitaine," *La Grisette: Revue populaire illustrée*, October 10, 1895, 83–85.

29. See the advertisement from the Office des Inventions Réunies in Album 7 mentioned in note 27.

30. This conversion is based off of information provided at https://www.historical statistics.org/Currencyconverter.html.

31. An example of an advertisement offering dildos at different sizes for a variety of prices can be found in Album 7 of the Milford Haven collection at the British Library. This text comes from a three-page fold-out advertisement, on a page labeled *godmiché* (dildo), with no information about the seller.

32. Edmond Picard, *La Veillée de L'huissier: Conte de Noël* (Brussels: Ferdinand Larcier, 1885), no page numbers.

33. *Matchett's Baltimore Director for 1851* (Baltimore: R. J. Matchett, 1851), 63, 216.

34. William Henry Boyd, *The Baltimore City Directory* (Baltimore: R. Edwards and W. H. Boyd, 1858), 127, 328; *Baltimore City Business Directory for 1858–'59* (Baltimore: J. C. Nicholson and H. Q. Nicholson, 1858), 60.

35. If anything, rubber manufacturing and distribution seems to have been far more prevalent in the New York City area. For example, an exhibit catalog from the Third Annual Exhibition of the Maryland Institute for the Promotion of the Mechanic Arts, held in Baltimore in 1850, notes that "gum elastic" items were contributed by a number of companies located in other states. These items included rubber belting from the New York company Reece & Hoyt, general rubber goods from a company called Union India Rubber Company, also in New York, and a collection of "nine dozen men, women and children's metallic boots, shoes and sandals, from the Goodyear Manufacturing Rubber Shoe Company" located in Connecticut. *Third Annual Exhibition of the Maryland Institute for the Promotion of the Mechanic Arts* (Baltimore: Sherwood & Co., 1850), 1, 11, 9. Indeed, a New York City directory for the year 1854–1855 includes no fewer than nineteen listings for businesses operating in India rubber, as well as numerous additional companies with names like New Brunswick Rubber Company, New York India Rubber Warehouse, Goodyear's Rubber Packing Co., and Goodyear Rubber Emporium. Charles R. Rode, *New-York City Directory, for 1854–1855* (New York: Doggett and Rode, 1854), 285. One thing you realize quickly in trying to make sense of the mid-nineteenth-century rubber industry in the United States is that business owners with no ostensible connection to Charles Goodyear seemed to have zero qualms about branding their companies with his name.

36. *Third Annual Exhibition of the Maryland Institute.*

37. These patents were issued in 1854 (patent number 11,135) and 1858 (patent number 22,080), respectively.

38. Société Excelsior, *Catalogue général d'articles de preservation intime à l'usage des deux sexes* (Paris: 1905), 54.

39. Mollier, *Le Camelot et la rue*, 7.

40. Basic information regarding the Paris World's Fairs has been drawn from material published by the curators of Gallica, an online archive associated with the Bibliothèque Nationale de France. See Catherine Brial, "Les expositions universelles dans Gallica," *Le Blog Gallica*, January 1, 2013, https://gallica.bnf.fr/blog/01012013 /les-expositions-universelles-dans-gallica.

41. For examples of historical scholarship illustrating the centrality of technology at the world's fairs and their importance for the development of twenty-first-century technologies, see Ron Becker, "'Hear-and-See Radio' in the World of Tomorrow: RCA and the Presentation of Television at the World's Fair, 1939–1940," *Historical Journal of Film, Radio and Television* 21, no. 4 (October 2001): 361–378; Paul Mason Fotsch, "The Building of a Superhighway Future at the New York World's Fair," *Cultural Critique* 48, no. 1 (2001): 65–97.

42. The 1900 Paris World's Fair alone reportedly drew fifty million visitors. Mollier, *Le Camelot et la rue*, 24.

43. Robert W. Rydell, "'Darkest Africa': African Shows at America's World's Fairs, 1893–1940," in *Africans on Stage: Studies in Ethnological Show Business*, ed. Bernth Lindfors (Bloomington: Indiana University Press, 1999), 135–155; Nathan Cardon, *A Dream of the Future: Race, Empire, and Modernity at the Atlanta and Nashville World's Fairs* (New York: Oxford University Press, 2018).

44. Isabel Morais, "'Little Black Rose' at the 1934 *Exposicao Colonial Portuguesa*," in *Gendering the Fair: Histories of Women and Gender at World's Fairs*, ed. Tracey Jean Boisseau and Abigail M. Markwyn (Urbana: University of Illinois Press, 2010), 21.

45. Walter Putnam, "'Please Don't Feed the Natives': Human Zoos, Colonial Desire, and Bodies on Display," *French Literature Series* 39 (2012): 55–68; Sadiah Qureshi, "Displaying Sara Baartman, the 'Hottentot Venus,'" *History of Science* 42, no. 2 (2004): 233–257.

46. Tracey Jean Boisseau and Abigail M. Markwyn, eds., *Gendering the Fair: Histories of Women and Gender at World's Fairs* (Urbana: University of Illinois Press, 2010).

47. Cheryl R. Ganz, *The 1933 Chicago World's Fair: A Century of Progress* (Urbana: University of Illinois Press, 2012), 22.

48. Henri de la Madelène, "Figaro à l'exposition," *Le Figaro*, September 16, 1855, 7.

49. de la Madelène, "Figaro à l'exposition," 7; my original English translation. Note that the ellipses in this passage are present in the original; they do not indicate missing text.

50. de la Madelène, 7.

51. Étienne Ducret, "Le Caoutchouc," *La Chanson: Journal de musique populaire*, May 15, 1880, 8. It appears that this 1880 version of "Le Caoutchouc" may actually

have been a rewrite of an earlier song by the same title. A listing for a song by the same title appears in the Parisian circular *Lice Chansonnière* in 1843. Unfortunately, though the circular is cataloged in Gallica, it appears its contents have been lost. *Lice Chansonnière* (Paris: L. Vieillot, 1843).

52. Ducret, "Le Caoutchouc," 8.

53. Ducret, 8. A "finger cot" (*doigtier*) was a rubber finger cover.

54. Étienne Ducret, "Le Caoutchouc," *Lice Chansonnière* (Paris: no publisher listed, 1898), 121–125. The listed authors of both the 1880 and 1898 versions of the song are the same, but it is unclear whether Étienne Ducret was indeed involved in the writing of both versions or whether the songs are being credited to him as the original author without note of who has updated them.

55. "Notes sur l'exposition," *La Vie Parisienne*, August 3, 1867, 558.

56. "Notes sur l'exposition," *La Vie Parisienne*, August 17, 1867, 592.

57. "Les Femmes en caoutchouc."

58. George Méliès (dir.), *L'Homme à la tête en caoutchouc* (Star Film Company, 1901).

59. Alphonse Lafitte, "La Femme en caoutchouc," *La Tintamarre*, March 10, 1872, 2–3.

60. Lafitte, "La Femme en caoutchouc," 3.

61. Rexer, *Fallen Veil*, 26.

62. "La Fraude," *Le Radical*, August 15, 1887, 2; Jean Frollo, "L'Imagination des fraudeurs," *Le Petit Parisien*, September 17, 1890, 1; Georges Acker, "Les Contrabandiers," *La Lanterne*, January 6, 1897, 2.

63. Frollo, "L'Imagination des fraudeurs," 1; Acker, "Les Contrabandiers," 2.

64. Acker, "Les Contrabandiers," 2.

65. Acker, 2.

66. Mollier, *Le Camelot et la rue*, 8.

67. Mollier, 244.

68. Mollier, 260.

69. Mollier, 148–149.

70. Mollier, 261.

71. F. Bellay, "Un Traquenard policier," *L'intransigeant*, October 24, 1900, 2; "Le Cas de l'empereur des camelots," *La Justice*, October 24, 1900, 2.

72. Bellay, "Un Traquenard policier," 2.

73. "Le Cas de l'empereur des camelots," 2.

74. "Le Cas de l'empereur des camelots," 2.

75. M. Pierre, "Carnet Judiciare: Femmes en caoutchouc!," *Gil Blas*, January 19, 1902, 3; F. Bellay, "La Traite des . . . poupées," *L'intransigeant*, January 20, 1902, 2; "Nouvelles judicaires," *Le Radical*, January 20, 1902, 2.

76. Bellay, "La Traite des . . . poupées," 2.

77. For example: Clément Voutel, "La Poupée," *La Vie Parisienne*, October 15, 1921, 883, 886–887.

78. Voutel, "La Poupée," 886.

79. This shift is illustrated in a 1919 catalog from Maison Claverie (which has seemingly rebranded as Établissements A. Claverie) for an artificial leg with the product name La Française. The catalog's title page includes both a note that Claverie is now an official supplier to military hospitals and a listing of Claverie's many awards from various world's fairs between 1912 and 1919. Here the remedicalization of rubber manufacture meets the long history of associating technological prowess with the world's fairs. Établissements A. Claverie, *La Jambe Artificielle "La Française"* (Paris: Établissements A. Claverie, 1919).

80. Lynn Comella, *Vibrator Nation: How Feminist Sex-Toy Stores Changed the Business of Pleasure* (Durham, NC: Duke University Press, 2017), 114.

81. Paisley Gilmour, "How to Tell If Your Sex Toy Is Toxic," *VICE*, March 1, 2018, https://www.vice.com/en/article/bj5bqv/how-to-tell-if-your-sex-toy-is-toxic; Dangerous Lilly, "Yes, Jelly Sex Toys Can Be Dangerous," *Dangerous Lilly* (blog), October 6, 2010, http://dangerouslilly.com/2010/10/yes-jelly-sex-toys-can-be-dangerous/.

82. Kim Airs, "Flesh and Fantasy: What's New in the Sex Doll Industry," *AVN*, January 22, 2020, https://avn.com/business/articles/novelty/flesh-and-fantasy-whats-new-in-the-sex-doll-industry-861528.html.

83. Amanda Phillips, "Dicks, Dicks, Dicks: Hardness and Flaccidity in (Virtual) Masculinity," *Flow: A Critical Forum on Media and Culture*, November 27, 2017, https://www.flowjournal.org/2017/11/dicks-dicks-dicks/.

Chapter 5

1. For writing on teledildonics, see Howard Rheingold, "Teledildonics and Beyond," in *The Postmodern Presence: Readings on Postmodernism in American Culture and Society* (Lanham, MD: AltaMira Press, 2005), 274–287; Teddy Pozo, "Haptic Media: Sexuality, Gender, and Affect in Technology Culture, 1959–2015" (PhD diss., University

of California, Santa Barbara, 2016); Maria Joao Faustino, "Rebooting an Old Script by New Means: Teledildonics—the Technological Return to the 'Coital Imperative,'" *Sexuality & Culture* 22, no. 1 (2018): 243–257.

2. For examples of Machulis's ongoing work on teledildonic butt plugs, see the portfolio of projects on his website (https://kyle.machul.is/portfolio/), as well as news reporting about his projects, such as: Samantha Cole, "This Animal Crossing–Enabled Buttplug Will Let You Hook Up In-Game," *VICE*, June 30, 2020, https://www.vice.com/en/article/pkyk9y/animal-crossing-connected-buttplug-vibrator.

3. Ruth Oldenziel, *Making Technology Masculine: Men, Women and Modern Machines in America, 1870–1945* (Amsterdam: Amsterdam University Press, 1999), 9.

4. Oldenziel, *Making Technology Masculine*, 9.

5. Oldenziel, 10.

6. Mar Hicks, *Programmed Inequality: How Britain Discarded Women Technologists and Lost Its Edge in Computing* (Cambridge, MA: MIT Press, 2017); Jennifer S. Light, "When Women Were Computers," *Technology and Culture* 40, no. 3 (1999): 455–483.

7. Laine Nooney, "The Uncredited: Work, Women, and the Making of the U.S. Computer Game Industry," *Feminist Media Histories* 6, no. 1 (January 1, 2020): 119–146; Adrienne Massanari, "#Gamergate and the Fappening: How Reddit's Algorithm, Governance, and Culture Support Toxic Technocultures," *New Media & Society* 19, no. 3 (March 2017): 329–346.

8. Joanna Radin, "Digital Dystopias: How Michael Crichton Taught Me to Start Worrying and Fear the Future," presented to Data & Society as part of the Future Perfect Conference, June 16, 2017; Sherry Turkle, *The Second Self: Computers and the Human Spirit* (Cambridge, MA: MIT Press, 2005), 200.

9. Julie Wosk, *My Fair Ladies: Female Robots, Androids, and Other Artificial Eves* (New Brunswick: Rutgers University Press, 2015), 6–7.

10. Rachel P. Maines, *The Technology of Orgasm: "Hysteria," the Vibrator, and Women's Sexual Satisfaction* (Baltimore: Johns Hopkins University Press, 2001); Hallie Lieberman, *Buzz: A Stimulating History of the Sex Toy* (New York: Pegasus Books, 2017); Lynn Comella, *Vibrator Nation: How Feminist Sex-Toy Stores Changed the Business of Pleasure* (Durham, NC: Duke University Press, 2017).

11. Donna J. Drucker, *Contraception: A Concise History* (Cambridge, MA: MIT Press, 2020); Claire Jones, *The Business of Birth Control: Contraception and Commerce in Britain before the Sexual Revolution* (Manchester: Manchester University Press, 2021).

12. For an example of popular histories of sex and sexual technologies that repeat the trope of the "ancient dildo," see Kate Devlin, *Turned On: Science, Sex and Robots* (London: Bloomsbury Sigma, 2018), 22.

13. Anne Marie Balsamo, *Technologies of the Gendered Body: Reading Cyborg Women* (Durham, NC: Duke University Press, 1996), 10.

14. Erich Wulffen, *Der Sexualverbrecher* (Berlin: Langenscheidt, 1910); Leo Schidrow-itz, *Sittengeschichte des Lasters die Kulturepochen und ihre Leidenschaften* (Vienna: Verlag für Kulturforschung, 1927), 214.

15. Wulffen, *Der Sexualverbrecher*, 299.

16. Iwan Bloch, *Das Sexualleben unserer Zeit in seinen Beziehungen zur modernen Kultur* (Berlin: Louis Marcus Veriagsbuchhandlung, 1907), 648; René Schwaeblé, *Les Détraquées de Paris: Étude de moeurs contemporaines* (Paris: H. Daragon, 1910), 85.

17. See, for example, Anthony Ferguson, *The Sex Doll: A History* (Jefferson, NC: McFarland & Company, 2010), 1.

18. Clément Voutel, "La Poupée," *La Vie Parisienne*, October 15, 1921, 886.

19. Madame B. (Alphonse Momas), *La Femme endormie* (Melbourne: J. Renold, 1899).

20. Rebecca Clark, "Gag Reflexes: Sex Doll Slapstick and Fran Ross's Oreo," *Post45*, January 22, 2020, https://post45.org/2020/01/gag-reflexes-sex-doll-slapstick-and-fran -rosss-oreo/.

21. Others have identified more nuanced cultural politics related to gender and sexuality in the film—arguing alternately, for example, that the film questions but ultimately renormativizes sexual identity or that the film successfully represents queer kinship structures and queers gender norms. Kate O'Neill, "Female Effigies and Performances of Desire: A Consideration of Identity Performance in *Lars and the Real Girl*," *FORUM: University of Edinburgh Postgraduate Journal of Culture & the Arts*, no. 6 (2008): 1–13; Claire Sisco King and Isaac West, "This Could Be the Place: Queer Acceptance in *Lars and the Real Girl*," *QED: A Journal in GLBTQ Worldmaking* 1, no. 3 (2014): 59–84.

22. Olivia Belton, "Metaphors of Patriarchy in *Orphan Black* and *Westworld*," *Feminist Media Studies* 20, no. 8 (November 16, 2020): 1211–1225.

23. Anastasia Salter and Bridget Blodgett, *Toxic Geek Masculinity in Media* (Cham, Switzerland: Palgrave Macmillan, 2017).

24. Katherine Cross, "Press F to Revolt," in *Diversifying Barbie and Mortal Kombat: Intersectional Perspectives and Inclusive Designs in Gaming*, ed. Yasmin B. Kafai, Gabriela T. Richard, and Brendesha M. Tynes (Pittsburgh: ETC Press, 2016), 23–34.

25. Henri de la Madelène, "Figaro à l'exposition," *Le Figaro*, September 16, 1855, 7.

26. Alphonse Lafitte, "La Femme en caoutchouc," *Le Tintamarre*, March 10, 1872, 2–3; Georges Eclar, "La Femme du capitaine," *La Grisette: Revue populaire illustrée*, October 10, 1895, 83–85.

27. Lafitte, "La Femme en caoutchouc," 3; Eclar, "La Femme du capitaine," 85.

28. "Les Femmes en caoutchouc," *L'Indépendent de Mascara*, August 27, 1885, 3.

29. "Les Femmes en caoutchouc," 3.

30. "Les Femmes en caoutchouc," 3.

31. "Les Femmes en caoutchouc," 3.

32. Emma Grey Ellis, "Whitney Cummings—and Her Sex Robot—Take on Modern Womanhood," *WIRED*, July 31, 2019, https://www.wired.com/story/whitney-cummings-netflix-special/; Brittany Knuper, "It's about Time: *Dragula* Winner Landon Cider and the History of Drag Kings," The Mary Sue, November 4, 2019, https://www.themarysue.com/dragula-landon-cider-drag-king-winner/.

33. Initially, the satirical cartoon from 1868 of the rubber woman wearing Greco-Roman robes and inflating men made out of animal-skin condom material that I discuss in chapter 4 seems like a potential exception as its artist is credited as Alma Tadema. However, upon further research, Alma Tadema is not the name of a woman artist but rather an unhyphenated version of the name Lawrence Alma-Tadema, a Dutch painter who lived in England and was best known for his paintings depicting classical scenes. The joke here in "attributing" the comic to Alma Tadema is that it's an image, a distinctly ribald one, that has been drawn in the grandiose style of an Alma-Tadema painting. Image associated with the feature "Promenade au salon," *Journal Amusant*, May 16, 1868, 4–6, image on page 5. For information regarding Lawrence Alma-Tadema, see Louise Lippincott, *Lawrence Alma Tadema: Spring* (Malibu: J. Paul Getty Museum, 1990).

34. Ferguson, *Sex Doll*, 1.

35. See, for example, Ferguson, 16.

36. Bloch, *Beiträge zur Aetiologie der Psychopathia Sexualis*, 301.

37. Schwaeblé, *Les Détraquées de Paris*.

38. Madame B., *La Femme endormie*.

39. See, for example, Ferguson, *Sex Doll*, 16.

40. Jean Genet, *Querelle de Brest* (Paris: Gallimard, 2010).

41. Martii Lahti, "Dressing Up in Power: Tom of Finland and Gay Male Body Politics," *Journal of Homosexuality* 35, no. 3–4 (June 4, 1998): 185–205; Andrew Stephenson, "'Our Jolly Marine Wear': The Queer Fashionability of the Sailor Uniform in Interwar France and Britain," *Fashion, Style & Popular Culture* 3, no. 2 (March 1, 2016): 157–172.

42. Jack King, "The Gay Ecstasy of The Village People," BBC, August 4, 2020, https://www.bbc.com/culture/article/20200804-the-gay-ecstasy-of-the-village-people.

43. George Chauncey, *Gay New York: Gender, Urban Culture, and the Making of the Gay Male World* (New York: Basic Books, 2019).

44. In addition to the examples of queer studies scholarship mentioned earlier in this book, see Matthew Knip, "Homosocial Desire and Erotic Communitas in Melville's Imaginary: The Evidence of Van Buskirk," *ESQ: A Journal of Nineteenth-Century American Literature and Culture* 62, no. 2 (2016): 355–414; Kellen Bolt, "Squeezing Sperm: Nativism, Queer Contact, and the Futures of Democratic Intimacy in Moby-Dick," *ESQ: A Journal of Nineteenth-Century American Literature and Culture* 65, no. 2 (2019): 293–329.

45. B. R. Burg, *Sodomy and the Pirate Tradition: English Sea Rovers in the Seventeenth-Century Caribbean* (New York: New York University Press, 1995), xxvi, 108, 103.

46. Hans Turley, *Rum, Sodomy, and the Lash: Piracy, Sexuality, and Masculine Identity* (New York: New York University Press, 1999), 2.

47. Turley, *Rum, Sodomy, and the Lash*, 2, 9.

48. Stefan Helmreich, "The Genders of Waves," *Women's Studies Quarterly* 45, no. 1–2 (2017): 29.

49. For more on rethinking technology through notions of the sea, see my discussion of Melody Jue's *Wild Blue Media* and the "blue humanities" in the conclusion to this book. Melody Jue, *Wild Blue Media: Thinking through Seawater* (Durham, NC: Duke University Press, 2020).

50. Sebastian Zulch, "Embracing Radical Softness: An Interview with Poet and Artist Lora Mathis," HelloFlo, December 5, 2016, https://helloflo.com/embracing-radical-softness-interview-poet-artist-lora-mathis/. For more on radical softness in relation to technology and digital media, see Teddy Pozo, "Queer Games After Empathy: Feminism and Haptic Game Design Aesthetics from Consent to Cuteness to the Radically Soft," *Game Studies* 18, no. 3 (2018); Andi Schwartz, "Soft Femme Theory: Femme Internet Aesthetics and the Politics of 'Softness,'" *Social Media + Society* 6, no. 4 (October 2020).

51. Sadie Plant, "The Future Looms: Weaving Women and Cybernetics," *Body & Society* 1, no. 3–4 (1995): 45–64.

52. See, for example, Sarah Fox, Rachel Rose Ulgado, and Daniela Rosner, "Hacking Culture, Not Devices: Access and Recognition in Feminist Hackerspaces," in *Proceedings of the 18th ACM Conference on Computer Supported Cooperative Work & Social Computing* (Vancouver: ACM, 2015), 56–68.

53. Andi Schwartz, "Low Femme, Low Theory: An Ethno-Archive of Femme Internet Culture" (PhD diss., York University, 2020), 95.

54. Rhea Ashley Hoskin, "Femme Theory: Refocusing the Intersectional Lens," *Atlantis: Critical Studies in Gender, Culture & Social Justice* 38, no. 1 (2017): 95–109; Laura Brightwell and Allison Taylor, "Femme Theory: Refocusing the Intersectional Lens," *Journal of Lesbian Studies* 25, no. 1 (January 2, 2021): 18–35.

55. Andi Schwartz, "Locating Femme Theory Online," *First Monday*, July 1, 2018, https://firstmonday.org/ojs/index.php/fm/article/view/9266.

56. Gabrielle Kassel, "Being a 'Queer Femme' Is about More than Just the Way You Dress," *Women's Health*, July 16, 2020, https://www.womenshealthmag.com/relation ships/a33299024/queer-femme/.

Chapter 6

1. One particularly unexpected node in the cultural constellation that surrounds the story of the Dutch wives is American artist Jasper Johns's two visual works titled *The Dutch Wives*, one an encaustic painting with collage elements from 1975 and the other a screen print from 1977. Both of Johns's pieces are abstract works featuring a crosshatch pattern disrupted by a small dark spot circled in red. Art historians have passed down their own lore about the works' title, with many (most notably Michael Crichton, writing in 1994) repeating a statement that a "Dutch wife" was a wooden "board with a hole, used by sailors as a surrogate for a woman"—suggesting that the dark spot that appears in Johns's works represents a hole for insertion. However, this explanation is itself suspect. In *Jasper Johns: Gray*, James Rondeau et al. give a commendably skeptical and thorough footnote documenting the lineage behind this backstory, where they write: "[Crichton] did not attribute the definition [of *Dutch wife*] to the artist in his published text, and we cannot substitute it in any other source," explaining that Crichton later claimed that Johns himself had given this definition for the term in a series of interviews conducted in 1976 and 1977. The authors go on to state that "the phrase is one of many in the English language that uses Dutch as a shorthand for a derogatory characterization. . . . The phrase 'Dutch wife' can also be used to refer to an open-frame bolster bed, made of bamboo or thick rattan . . . It is possible that the title refers to the formal qualities of such a frame as related to the crosshatch pattern." James Rondeau, Jasper Johns, Douglas W. Druick, Mark Pascale, and Nan Rosenthal, *Jasper Johns: Gray* (Chicago: Art Institute of Chicago, 2008), 77; Michael Crichton, *Jasper Johns* (New York: Harry N. Abrams, Inc., 1994).

2. For an example of a text that says the Dutch wives were sex dolls made by Asian people and adopted by Dutch sailors, see Julian Arbois, *Dans le lit de nos ancêtres: Sexualité, moeurs, et vie intime d'autrefois* (Bernay, France: City Editions, 2016), 212.

3. Kerry Ward, *Networks of Empire: Forced Migration in the Dutch East India Company* (Cambridge: Cambridge University Press, 2009).

4. For an example of a text that says that Dutch wives were sex dolls originally made by Dutch sailors and brought to Asia, see Anthony Ferguson, *The Sex Doll: A History* (Jefferson, NC: McFarland & Company, 2010), 27.

5. For an example of a text that says that the Dutch wives were sex dolls made by Dutch sailors after seeing bamboo sleeping cages in Asia, see Priscille Lamure, "Petite histoire de la poupée érotique," *Savoirs d'Histoire* (blog), October 26, 2017, https:// savoirsdhistoire.wordpress.com/2017/10/26/petite-histoire-de-la-poupee-erotique /comment-page-1/.

6. Most contemporary tellings of this story ultimately point back to Alan Scott Pate's book *Ningyo: The Art of the Japanese Doll* (North Clarendon, VT: Tuttle Publishing, 2005), which includes a description of elaborate sex dolls supposedly made in eighteenth-century Osaka. This description has similarities to certain accounts of the dames de voyage (the dolls are mechanized, warm water can be poured into the dolls, they are referred to as "traveling beauties," etc.). However, Pate admits that he has been unable to find any surviving material evidence of these sex dolls (275). He gives no citation for his statements about the traveling beauties, though David Levy seems to have contacted Pate to ask for his source (Levy, "Intimate Relationships with Artificial Partners" [PhD diss., Maastricht University, 2007], v). Pate, in turn, claims that he drew his info from Mitamura Engyo's text "Takeda Hachidai" or "Eight Generations of the Takeda Family" (Levy, *Love and Sex with Robots: The Evolution of Human-Robot Relations* [New York: Harper Collins, 2007], 249). However, "Eight Generations of the Takeda Family" is not a primary account from either the 1600s or 1700s. Rather, it is a work from the first half of the twentieth century, part of a larger oeuvre about the morals and customs of everyday life in Edo-era Japan. Although I have not been able to access Engyo's text in full, my sense is that it doesn't tell a version of the story of the traveling beauties that fully matches Pate's. Even if it did, it's equally possible that the story was an early twentieth-century invention. All of this matches up with Japanese historian Agnès Giard's assertion that contemporary accounts of early Japanese sex dolls are erroneous. Mitamura Engyo, *Takeda Hachidai*, vol. 21 of *Mitamura Engyo Zenshu* [*The Complete Works of Engyo Mitamura*] (Tokyo: Chuo Koronsha, 1976); the original date of publication is unclear, but the author lived from 1870 to 1952. Agnès Giard, *Un désir d'humain: Les "love doll" au Japon* (Paris: Les Belles Lettres, 2016), 40.

7. There are a number of ways to walk through this citational lineage, but one illustrative trail looks like this: Priscille Lamure's online article "Petite histoire de la poupée érotique," which has since been picked up by many other sources, talks about the Dutch wives as sex dolls (and also ties them directly to the dames de voyage). Lamure claims to be drawing from Julien Arbois's book *Dans le lit de nos ancêtres: Sexualité, moeurs, et vie intime d'autrefois*, which in turn claims to be drawing

from Agnès Giard, "Pourquoi les poupées gonflables sappellent des epouses hollandaises" [Why inflatable dolls are called Dutch wives], *Libération*, January 23, 2013, https://www.liberation.fr/debats/2013/01/23/pourquoi-les-poupees-gonflables-s -appellent-des-epouses-neerlandaises_1811808/. Yet as information moves between these texts, we see a number of slippages. For example, Arbois misses the nuance of Giard's article, where she describes how bamboo sleeping cages filled a symbolic role in cultural traditions involving husbands and wives, and instead reports that the "bamboo wives" were actual, literal sex dolls. There are also a number of claims in Arbois's text that clearly do not come from Giard, but rather from Anthony Ferguson's uncited *The Sex Doll: A History*, or another intermediary text through which Arbois encounters Ferguson's work, though Arbois again scrambles Ferguson's claims. To support his own claims that seventeenth-century European merchant ships brought leather sex dolls to Japan through the Dutch East India Company, Ferguson points (in a roundabout way) to David Levy's *Love and Sex with Robots*. Levy cites Alan Scott Pate's *Ningyo: The Art of the Japanese Doll*, in which Pate writes that contemporary sex dolls in Japan are called "Dutch wives" because Dutch merchant marine vessels carried them onboard (Pate, 275). Unlike Ferguson or Levy, however, Pate then expands on this story by describing how Dutch wives were supposedly produced in Japan itself a hundred years later: "In the eighteenth century, the Yamamoto and Takeda families in Osaka were known for the creation of Japanese versions of 'Dutch wives,' called *shutsuro bijin* (lit. traveling beauties) or *koshoku onna* (play women). Highly realistic in their execution, these dolls incorporated a device that allowed warm water to be poured inside the figure, giving it a greater verisimilitude, as well as zenmai spring-driven mechanisms that allowed their arms and legs to be moved. Exceptionally popular among government officials stationed in Osaka, they were banned during the Horeki era (1751–63)" (275). What we see here is how the story of the Dutch wives has moved through a very similar trajectory of citation, missing citation, adaptation, and so on as the tale of the dames de voyage.

8. Giard, "Pourquoi les poupées gonflables."

9. Mario Esposito, "What's a Bamboo Wife?," Good Night's Rest, June 24, 2019, https://www.goodnights.rest/about-pillows-bolsters-cushions/bamboo-wife/.

10. Giard, "Pourquoi les poupées gonflables."

11. Hugh Wilkson, "1882: A Dutch Wife," in *Travellers' Tales of Old Singapore*, compiled by Michael Wise (Singapore: Marshall Cavendish Editions, 1998), no printed page numbers.

12. An example illustrating this sense that the bamboo sleeping cages come from an earlier historical time can be found in the 1985 Korean short story "My Tale of the Bamboo Wife," discussed in greater length ahead, in which a younger character finds a "bamboo wife" and brings it to an older character to ask what it is, explaining,

"Considering your venerable age, sir, I thought you might know." Hwang Sun-wŏn, "My Tale of the Bamboo Wife," trans. Bruce Fulton and Ju-Chan Fulton, *Azalea: Journal of Korean Literature & Culture* 8, no. 1 (2015): 174.

13. For more on Japanese obscenity laws and their relationship to sex toys, see my discussion of the rabbit vibrator in chapter 3.

14. Lamure, "Petite Histoire."

15. For more on connections between prostitution and port cities, see my discussion of sex work in relation to maritime studies in chapter 3.

16. Wilkson, "1882: A Dutch Wife."

17. Wilkson.

18. Wilkson.

19. "A Sleepless Night in the Tropics," *Ballou's Monthly Magazine* 75, no. 6 (June 1892): 462–464.

20. "A Sleepless Night in the Tropics," 462.

21. "A Sleepless Night in the Tropics," 463.

22. Hwang Sun-wŏn, "My Tale of the Bamboo Wife."

23. Hwang Sun-wŏn, 178.

24. Hwang Sun-wŏn, 179.

25. Some examples of texts discussed elsewhere in this book that include sexual technologies from Asia in their histories are Paul Tabori, *The Humor and Technology of Sex* (New York: Julian Press, 1969); Ferguson, *Sex Doll*; Levy, *Love and Sex with Robots*.

26. Carlin Wing, "Episodes in the Life of Bounce: Playing with a Rubber Ball," *Cabinet*, no. 56 (2014–2015), https://www.cabinetmagazine.org/issues/56/wing.php.

27. Anjali R. Arondekar, *For the Record: On Sexuality and the Colonial Archive in India* (Durham, NC: Duke University Press, 2009), 100; Marquis de Sade, *Philosophy in the Boudoir or, The Immoral Mentors*, trans. Joachim Neugroschel (New York: Penguin Books, 2006). For more on dildos in *Philosophy in the Boudoir*, see Liza Blake, "Dildos and Accessories: The Functions of Early Modern Strap-Ons," in *Ornamentalism: The Art of Renaissance Accessories*, ed. Bella Mirabella (Ann Arbor: University of Michigan Press, 2011), 130–155.

28. Arondekar, *For the Record*, 97–98.

29. Clément Voutel, "La Poupée," *La Vie Parisienne*, October 15, 1921, 883, 886–887.

30. "Les Femmes en caoutchouc," *L'Indépendent de Mascara*, August 27, 1885, 3.

31. Henri de la Madelène, "Figaro à l'exposition," *Le Figaro*, September 16, 1855, 7.

32. Louis Chude-Sokei, "The Uncanny History of Minstrels and Machines, 1835–1923," in *Burnt Cork: Traditions and Legacies of Blackface Minstrelsy*, ed. Stephen Johnson (Amherst: University of Massachusetts Press, 2012), 104–132; Catherine A. Stewart, *Long Past Slavery: Representing Race in the Federal Writers' Project* (Chapel Hill: University of North Carolina Press, 2016), 13.

33. Voutel, "La Poupée," 886.

34. Moya Bailey, *Misogynoir Transformed: Black Women's Digital Resistance* (New York: New York University Press, 2021).

35. Maison L. Bador, *Fabrication perfectionée de caoutchouc dilaté et baudruche* (Paris: commercial catalog, 1900), 4.

36. Maison L. Bador, *Fabrication perfectionée*, 4.

37. Jessica Borge, *Protective Practices: A History of the London Rubber Company and the Condom Business* (Montreal: McGill-Queen's University Press, 2020), 16.

38. Advertisement (center of page), *La Grisette: Revue populaire illustrée*, August 10, 1895, 2.

39. Advertisement, *La Grisette*.

40. For an example of an account that emphasizes the idea that the dames de voyage were stored in ships' holds, see Arbois, *Dans le lit de nos ancêtres*, 213.

41. Sowande' M. Mustakeem, *Slavery at Sea: Terror, Sex, and Sickness in the Middle Passage* (Urbana: University of Illinois Press, 2016), 9.

42. Mustakeem, *Slavery at Sea*, 84.

43. Mustakeem, 86–87.

44. Chunghee Sarah Soh, *The Comfort Women: Sexual Violence and Postcolonial Memory in Korea and Japan* (Chicago: University of Chicago Press, 2008).

45. Madame B., *La Femme endormie*.

46. For a particularly recent example of yet another popular history that points back to *La Femme endormie*, see Kate Lister, *A Curious History of Sex* (London: Unbound, 2020).

47. Madame B., *La Femme endormie*.

48. Alexandre Dumas, *Captain Pamphile*, trans. Andrew Brown (London: Hesperus Classics, 2006).

49. Daniel Defoe, *Robinson Crusoe* (New York: Penguin Classics, 2006); Herman Melville, *Moby-Dick* (New York: Macmillan, 2016); J. K. Huysmans, *Against Nature* (New York: Penguin Books, 2003).

50. Andrew Brown, "Introduction," in Alexandre Dumas, *Captain Pamphile*, trans. Andrew Brown (London: Hesperus Classics, 2006), xi–xv, xi.

51. Alexandre Dumas, *The Count of Monte Cristo* (New York: Penguin Books, 2003); Dumas, *The Three Musketeers* (New York: Penguin Books, 1982).

52. Dumas, *Captain Pamphile*, 88.

53. Dumas, *Captain Pamphile*, 140, 150, 151.

54. Dumas, *Captain Pamphile*, 153.

55. Eric Martone, "Introduction: Alexandre Dumas as a Francophone Writer," in *The Black Musketeer: Reevaluating Alexandre Dumas within the Francophone World*, ed. Eric Martone (Newcastle upon Tyne: Cambridge Scholars Publishing, 2011), 1–32, 2.

56. Madame B., *La Femme endormie.*

57. Neda Atanasoski and Kalindi Vora, *Surrogate Humanity: Race, Robots, and the Politics of Technological Futures* (Durham, NC: Duke University Press, 2019), 190, 192.

58. Atanasoski and Vora, *Surrogate Humanity*, 193.

59. Atanasoski and Vora, 193.

60. Nettrice R. Gaskins, "Techno-Vernacular Creativity and Innovation across the African Diaspora and Global South," in *Captivating Technology: Race, Carceral Technoscience, and Liberatory Imagination in Everyday Life*, ed. Ruha Benjamin (Durham, NC: Duke University Pres, 2019), 253–274.

61. Gaskins, "Techno-Vernacular Creativity," 252.

62. Gaskins, 252, 253.

63. Ruha Benjamin, "Introduction: Discriminatory Design, Liberating Imagination," in *Captivating Technology: Race, Carceral Technoscience, and Liberatory Imagination in Everyday Life*, ed. Ruha Benjamin (Durham, NC: Duke University Pres, 2019), 1–22, 12.

64. Mustakeem, *Slavery at Sea*, 125.

65. Jessica Marie Johnson, "Markup Bodies," *Social Text* 36, no. 4 (December 1, 2018): 57–79.

66. M. Oliver, "Lupe Fiasco's 'Drogas Wave' Could Have Been a Great Album with a Bit of Editing," *Pop Matters*, October 1, 2018, https://www.popmatters.com/lupe-fiasco-drogas-wave-2609079350.html.

67. Nettrice R. Gaskins, "Deep Sea Dwellers: Drexciya and the Sonic Third Space," *Shima: The International Journal of Research into Island Cultures* 10, no. 2 (October 10, 2016), 68–80: 68, 73, 72.

68. Ytasha Womack, *Afrofuturism: The World of Black Sci-Fi and Fantasy Culture* (Chicago: Chicago Review Press, 2013), 9.

69. Womack, *Afrofuturism*, 9.

70. Suzanna Chan, "'Alive . . . Again.' Unmoored in the Aquafuture of Ellen Gallagher's Watery Ecstatic," *Women's Studies Quarterly* 45, no. 1–2 (2017): 246.

71. Alexis Pauline Gumbs, *Undrowned: Black Feminist Lessons from Marine Mammals* (Chico, CA: AK Press, 2020).

72. Gumbs, *Undrowned*, 2.

Chapter 7

1. Allison de Fren, "Technofetishism and the Uncanny Desires of A.S.F.R. (alt.sex. fetish.robots)," *Science Fiction Studies* 36, no. 3 (2009): 401–440, 407.

2. de Fren, "Technofetishism," 409.

3. Heather Schoenfeld, *Building the Prison State: Race and the Politics of Mass Incarceration* (Chicago: University of Chicago Press, 2018).

4. John Mercer, "In the Slammer: The Myth of the Prison in American Gay Pornographic Video," *Journal of Homosexuality* 47, no. 3–4 (September 15, 2004): 151–166.

5. David Levy, *Love and Sex with Robots: The Evolution of Human-Robot Relations* (New York: Harper Collins, 2007), 181.

6. Cynthia Ann Moya, "Artificial Vaginas and Sex Dolls: An Erotological Investigation" (PhD diss., Institute for Advanced Study of Human Sexuality, 2006), 44.

7. Leo Schidrowitz, *Ergänzungswerk zur Sittengeschichte des Lasters* (Vienna: Verlag für Kulturforschung, 1927).

8. Schidrowitz, *Ergänzungswerk*, no page numbers, figure XX.

9. Making sense of the various published versions of Hirschfeld's *Sexualpathologie* is a bizarrely vertiginous business as the work contains multiple volumes that were published out of order and were then republished together in a later volume. The text in which this image appears is the third volume of three, which was originally published in 1920. The page numbers I give for this text are internal to this volume specifically. This is helpful to know for others looking for this image because the 1921 full version of *Sexualpathologie*, which has all three volumes, has different sets

of page numbers for each volume. Magnus Hirschfeld, *Sexualpathologie: Ein Lehrbuch für Ärtze und Studierende: Dritter Teil* (Bonn: A. Marcus & E. Webers Verlag, 1920). Image appears on unnumbered page between pages 128 and 129.

10. Hirschfeld, *Sexualpathologie*, 129.

11. Hirschfeld, unnumbered image page between pages 128 and 129.

12. Angelo, *Prisoners' Inventions* (Chicago: Whitewalls, 2003).

13. Angelo, *Prisoners' Inventions*.

14. Angelo.

15. Angelo.

16. Angelo.

17. Rebecca Clark, "Gag Reflexes: Sex Doll Slapstick and Fran Ross's Oreo," *Post45*, January 22, 2020, https://post45.org/2020/01/gag-reflexes-sex-doll-slapstick-and-fran -rosss-oreo/.

18. Anthony Ferguson, *The Sex Doll: A History* (Jefferson, NC: McFarland & Company, 2010).

19. Ferguson, *Sex Doll*, 24–25.

20. Graeme Donald, *Mussolini's Barber: And Other Stories of the Unknown Players Who Made History Happen* (Oxford: Osprey, 2010).

21. "Hitler Ordered Nazis to Make Sex Dolls So Soldiers Wouldn't Catch Syphilis from Prostitutes," *Daily Mail*, July 11, 2011, https://www.dailymail.co.uk/news /-2013397/Hitler-ordered-Nazis-make-sex-dolls-soldiers-wouldnt-catch-syphilis -prostitutes.html.

22. Ferguson, *Sex Doll*, 25.

23. One point of connection to the lineage of the tale of the dames de voyage is the anecdote from the pseudoscientific book *Sex Devices and How to Use Them*, in which a man describes seeing a live sex show featuring a sex doll during his time as an American soldier in Germany. This suggests that the story of the Nazi blow-up doll likely has connections to earlier stories about sex dolls in relation to World War II. Amy Dumont and Ashton Dumont, *Sex Devices and How to Use Them* (Los Angeles: Argyle Books, 1970).

24. Nick Dyer-Witheford and Greig de Peuter, "Empire@play: Virtual Games and Global Capitalism," *CTheory* (2009): 5–13.

25. Clark, "Gag Reflexes"; Ferguson, *Sex Doll*, 24.

26. Bethania Palma, "Did Hitler Invent the Inflatable Sex Doll?," Snopes, July 22, 2020, https://www.snopes.com/fact-check/inflatable-sex-doll/; Stephen Johnson, "Did Hitler Really Invent the Sex Doll?," 13th Floor, December 6, 2016, http://www.the13thfloor.tv/2016/12/06/did-adolf-hitler-invent-the-sex-doll/ (accessed May 15, 2021); Ian Smith, "FACT or FICTION? Hitler Invented the Inflatable Sex Doll . . . Allegedly!," The Vintage News, September 14, 2015, https://www.thevintagenews.com/2015/09/14/fact-or-fiction-hitler-invented-the-inflatable-sex-doll-allegedly/.

27. Ferguson, *Sex Doll*, 25.

28. Ferguson, 25.

29. Levy, *Love and Sex with Robots*, 22.

30. Levy, 177.

31. Auguste Villiers de l'Isle-Adam, *L'Ève future* (Paris: Ancienne Maison Monnier, 1886); Madame B. (Alphonse Momas), *La Femme endormie* (Melbourne: J. Renold, 1899); René Schwaeblé, *Les Détraquées de Paris: Étude Documentaire* (Paris: Bibliothèque du Fin de siècle, 1904).

32. Iwan Bloch, *The Sexual Life of Our Time in Its Relations to Modern Civilization*, trans. M. Eden Paul (London: Rebman Limited, 1909).

33. Bloch, *Sexual Life of Our Time*, 648.

34. Ferguson, *Love Doll*, 11, 12.

35. Ovid, *Metamorphoses*, trans. Rolfe Humphries (Bloomington: Indiana University Press, 2018).

36. Ovid, *Metamorphoses*, 242.

37. Ovid, 242.

38. Ovid, 241.

39. Ovid, 240–241.

40. Ovid, 243.

41. Ovid, 243.

42. Ovid, 243.

43. Ovid, 243.

44. Ovid, 243.

45. Ovid, 242.

46. Bloch, *Sexual Life of Our Time*, 648.

47. Ovid, *Metamorphoses*, 242.

48. Carlo Collodi, *The Adventures of Pinocchio*, trans. Ann Lawson Lucas (Oxford: Oxford University Press, 1996).

Conclusion

1. Anaïs Nin, *Delta of Venus* (New York: Pocket Books, 1977).

2. Mary Wollstonecraft Shelley, *Frankenstein: The 1818 Text* (New York: Penguin Books, 2018).

3. Bliss Cua Lim, "Dolls in Fragments: Daisies as Feminist Allegory," *Camera Obscura* 16, no. 2 (2001): 37–77, 47, 51.

4. Julie Wosk, *My Fair Ladies: Female Robots, Androids, and Other Artificial Eves* (New Brunswick: Rutgers University Press, 2015), 166.

5. Wosk, *My Fair Ladies*, 176–180.

6. Wosk, 171.

7. Nin, *Delta of Venus*; Anaïs Nin, *Little Birds* (New York: Harcourt Brace Jovanovich, 1979).

8. Karen Brennan, "Anaïs Nin: Author(iz)ing the Erotic Body," *Genders* 14 (1992): 66–86; Anna Powell, "Heterotica: The 1000 Tiny Sexes of Anaïs Nin," in *Deleuze and Sex*, ed. Frida Beckman (Edinburgh: Edinburgh University Press, 2011), 50–68.

9. Nin, *Delta of Venus*, 10.

10. Nin, 13.

11. Nin, 15.

12. Nin, 10.

13. Nin, 15.

14. Nin, 23.

15. Nin, 17–18.

16. Nin, 18.

17. Nin, 16–17.

18. Nin, 18.

19. Nin, 18.

20. Nin, 18.

21. Nin, 18.

22. Mel Y. Chen, *Animacies: Biopolitics, Racial Mattering, and Queer Affect* (Durham, NC: Duke University Press, 2012); Anna Lowenhaupt Tsing, *The Mushroom at the End of the World: On the Possibility of Life in Capitalist Ruins* (Princeton, NJ: Princeton University Press, 2015); Donna J. Haraway, *Simians, Cyborgs, and Women: The Reinvention of Nature* (New York: Routledge, 1991).

23. Nin, *Delta of Venus*, 18.

24. Bo Ruberg, "Queer Game Physics: The Gendered and Sexual Implications of How Video Games Move," presentation to the Society of Cinema and Media Studies Conference, March 17, 2021.

25. Susan Stryker, "My Words to Victor Frankenstein above the Village of Chamounix: Performing Transgender Rage," *GLQ: A Journal of Lesbian and Gay Studies* 1, no. 3 (1994): 237–254.

26. Nicole Starosielski, *The Undersea Network* (Durham, NC: Duke University Press, 2015); Miriam Posner, "The Software that Shapes Workers' Lives," *New Yorker*, March 12, 2019, https://www.newyorker.com/science/elements/the-software-that -shapes-workers-lives.

27. Ghislain Thibault, "Streaming: A Media Hydrography of Televisual Flows," *VIEW: Journal of European Television History and Culture* 4, no. 7 (2015): 110–119.

28. Stacy Alaimo, "Introduction: Science Studies and the Blue Humanities," *Configurations* 27, no. 4 (2019): 429–432; Janine MacLeod, Cecilia Chen, and Astrida Neimanis, *Thinking with Water* (Montreal: McGill-Queens University Press, 2013).

29. Stefan Helmreich, "The Genders of Waves," *Women's Studies Quarterly* 45, no. 1–2 (2017): 29–51.

30. Sowande' M. Mustakeem, *Slavery at Sea: Terror, Sex, and Sickness in the Middle Passage* (Urbana: University of Illinois Press, 2016), 5.

31. Melody Jue, *Wild Blue Media: Thinking through Seawater* (Durham, NC: Duke University Press, 2020).

32. Jue, *Wild Blue Media*, 6.

33. Jue, xi.

34. Terri Gordon-Zolov and Amy Sodaro, "Introduction," *Women's Studies Quarterly* 45, no. 1–2 (2017): 12.

Index